CLOCKWORK COSMOS TIME'S EVOLUTION

CLOCKWORK COSMOS
TIME'S EVOLUTION

EHSAN SHEROY

Noble Publishing

CONTENTS

INDEX

Introduction

The unyielding entry of time has for quite some time been a significant mystery that has dazzled the human creative mind. From the constant ticking of seconds to the inestimable artful dance of divine bodies, the idea of time is woven into the actual texture of our reality. In the excellent embroidery of the universe, time arises as a baffling power that oversees the unfurling of occasions, forming the past, present, and future in a dance of enormous extents. As we leave on an excursion through the Precision Universe, we dig into the complex instruments that drive time's development, unwinding the strings of transient secrets that have interested masterminds and researchers all through the ages.

At the core of this investigation lies the essential inquiry: What is time? Is it a changeless stream streaming unendingly, or a pliable aspect formed by the actual idea of the real world? The solutions to these inquiries rise above the limits of reasoning, material science, and mysticism, bringing us into a multi-layered request that rises above the constraints of our comprehension. Time, in its substance, is both an enormous steady and a unique power, directing the rhythms of presence with an accuracy that evades our total cognizance.

To grasp the development of time, we should wander past the recognizable domains of our day to day routines and embrace the enormous viewpoint. The Accuracy Universe, with its many-sided gears and divine instruments, fills in as the stage for time's amazing exhibition. From the glorious breadth of worlds to the tiny dance of particles, the universe unfurls in an orchestra of interconnected occasions, every second resounding with the reverberations of the past and the preface to a questionable future.

As we look at the night sky, we witness the old light of far off stars, their iridescence conveying the engravings of ages a distant memory. The divine bodies, limited by the gravitational hug of infinite powers, follow unpredictable directions through the immense field of room. It is inside this divine dance that we track down the main murmurs of time's development — a story written in the language of gravity, relativity, and the enormous transaction of issue and energy.

The excursion into the profundities of time's advancement takes us to the beginning of the actual universe. In the cauldron of the Enormous detonation, the ensemble of creation unfurled, bringing forth the grandiose scene that would turn into the material for time's unfurling. The vast microwave foundation radiation, a weak reverberation of the early stage blast, murmurs mysteries about the universe's outset, welcoming us to unravel the grandiose code implanted in its old light.

As the universe extended and cooled, the early stage soup of particles blended into the main iotas, proclaiming the rise of another age. The infinite timetable walked forward, winding around the embroidered artwork of inestimable advancement.

The dance of universes, the development of stars and planets, and the astronomical patterns of birth and demise turned into the vast brushstrokes arranging the representation of time's determined walk.

The interaction among issue and energy, represented by the laws of material science, shaped the forms of the universe. Einstein's hypothesis of relativity introduced another comprehension of existence, uncovering the interconnected idea of the texture that ties the universe. Gravity, as the curve of spacetime, arose as the grandiose choreographer, organizing the developments of divine bodies with a gravitational style that rises above human creative mind.

However, the narrative of time's advancement stretches out past the enormous material to the domains of quantum vulnerability. In the subatomic domain, particles resist the old style ideas of determinism, participating in a dance of likelihood and un-usualness. Quantum entrapment, superposition, and the perplexing way of behaving of particles challenge our instinctive accept of the real world, welcoming us to rethink the actual underpinnings of time itself.

The investigation of time's development likewise drives us to the astronomical balances and deviations that shape the unfurling of occasions. The bolt of time, point-ing constantly from the past into what's in store, acquaints a feeling of directionality with the vast story. The secrets of entropy and the second law of thermodynamics enlighten the irreversibility of time, denoting the movement from request to scatter in the fabulous grandiose show.

As we explore the inestimable landscape, we experience divine peculiarities that act as grandiose watches. Pulsars, astronomical timekeepers with striking accuracy, throb in the immensity of room, their musical beats resounding across light-years. Dark openings, inestimable conundrums with gravitational ravenousness, twist the texture of spacetime, making worldly abnormalities that challenge how we might interpret time's temperament.

The human experience of time, laced with the vast story, acquaints an emotional as-pect with our investigation. Social, mental, and philosophical points of view shape our impression of time, leading to different accounts that mesh into the all-encompassing embroidered artwork of the Accuracy Universe. The brief snapshots of our lives, the reverberations of old civilizations, and the desires of people in the future unite in a transient mosaic that rises above individual presence.

As we navigate the enormous scene, we face the wildernesses of our insight and the tempting secrets that keep on evading our comprehension. Dim matter and dim energy, confounding parts ruling the astronomical spending plan, allure us to investigate the secret components of the universe. The idea of time at the quantum scale, the chance of different universes, and the vast destiny of our universe stay open inquiries that impel the wildernesses of logical investigation into unfamiliar region.

In the great story of Precision Universe, time's development arises as a multi-layered embroidery woven with the strings of vast history, quantum vulnerability, and the human experience. From the vast ensemble of creation to the perplexing domains of quantum reality, the investigation of time rises above disciplinary limits, welcoming us to leave on an excursion of scholarly disclosure and insightful reflection.

As we explore the tremendous span of the Accuracy Universe, we become people who jump through time, navigating the ages of the universe and witnessing the infinite secrets that unfurl across the grandiose course of events. The pinion wheels of time, complicatedly interlocked with the divine apparatus, welcome us to contemplate the idea of presence, the beginnings of the universe, and a definitive predetermination that anticipates the unfurling embroidery of time.

In this investigation of Precision Universe: Time's Development, we leave on a mission to unwind the secrets that encompass time, digging into the domains of cosmology, quantum material science, and the human experience. The excursion vows to be both a scholarly odyssey and a scrutinizing reflection on the idea of reality itself. As we peer into the enormous pit, we are constrained to pose significant inquiries about the idea of time, our position in the universe, and the getting through secrets that keep on coaxing us toward the boondocks of information.

Chapter 1

The Genesis of Time

In the tremendous region of the universe, where the embroidery of room winds around its many-sided designs, there exists a power that rises above the limits of presence itself. This power, strange and transcendent, is the beginning of time — an enormous ensemble that coordinates the dance of heavenly bodies and the unfurling of infinite ages.

At the beginning of creation, when the universe was nevertheless a twirling cauldron of early stage energy, the seeds of time were planted. The universe, in its early state, was a material ready to be painted with the brushstrokes of endlessness. Inside the grandiose cauldron, the major components of reality combined and consolidated, bringing forth the actual texture of time.

As the ensemble of creation unfurled, the universe extended, and matter consolidated into systems, stars, and planets. Time, be that as it may, was not a simple onlooker in this enormous display; it was the expert guide, directing the cadence and stream of presence. In the enormous suggestion, the beginning of time appeared as a key power, molding the fate of all that existed.

In the pot of stars, where atomic combination touched off the vast heaters, the beginning of time played out its most terrific developments. The existence pattern of stars turned into a demonstration of the inflexible entry of time. Stars, brought into the world from the enormous belly, moved across the heavenly stage, consuming brilliantly thriving prior to surrendering to the steady walk of time.

Supernovae, the grandiose blasts that noticeable the final breaths of enormous stars, released a downpour of energy and components into the infinite embroidery. These heavenly passings, however destructive, were fundamental for the beginning of time. The garbage from supernovae improved the universe with the structure blocks of life, making the unrefined components for the development of planets and, in the end, conscious creatures.

On one such planet, settled in the edges of a twisting world, the beginning of time took another structure. As life rose up out of the early stage oceans, the dance of development unfurled — a dance arranged by the hands of time. The ages passed, and life advanced from straightforward single-celled living beings to perplexing, multicellular animals. With every heartbeat, with every breath, the beat of time reverberated through the hallways of organic presence.

In the domain of cognizant creatures, time took on another importance. The familiarity with time turned into a main quality of conscious life. The recurrent rhythms of constantly, the evolving seasons, and the unavoidable walk towards death — every one of these became markers of the progression of time. Time, when a theoretical power, turned out to be personally woven into the texture of cognizant experience.

As civic establishments rose and fell, leaving behind the reverberations of their accomplishments and disappointments, the beginning of time proceeded with its persevering walk. The incredible pyramids of Egypt, the antiquated urban areas of Mesopotamia, and the heavenly domains of Rome and China — all took the stand concerning the mind boggling dance among time and human undertaking. The remnants of antiquated human advancements remained as landmarks to the fleetingness of mortal accomplishments despite the everlasting power of time.

In the cauldron of human cognizance, the journey to comprehend and gauge time came to fruition. Sundials, water tickers, and hourglasses turned out to be early instruments to evaluate the progression of time. However, time stayed a tricky idea, rising above the grip of human instruments. The philosophical personalities of old scholars contemplated the idea of time, discussing whether it was a flat out the real world or an emotional deception.

With the coming of logical request, the beginning of time entered another stage. The progressive thoughts of masterminds like Galileo and Newton established the groundwork for a quantitative comprehension of time.

Time turned into a boundary, an aspect that could be estimated and examined. The ticking of clocks and the swinging of pendulums turned into the metronome of the logical age, denoting the beat of a steadily propelling comprehension of the universe.

In the mid twentieth 100 years, the hypothesis of relativity, figured out by Albert Einstein, introduced a change in perspective in how we might interpret time. Time was as of now not a uniform and outright element; it was a dynamic and pliant aspect entwined with space. The texture of spacetime distorted within the sight of monstrous items, forming the direction of divine bodies and affecting the progression of time itself.

The beginning of time, when an inestimable power directing the dance of stars, presently uncovered itself as a crucial part of the actual texture of the universe. Time, as grasped from the perspective of relativity, turned into a powerful waterway, streaming and twisting because of the gravitational scene of the universe. The interaction among gravity and time opened new vistas of investigation, testing our assumptions and extending the skylines of human comprehension.

In the domain of quantum mechanics, one more wilderness of the enormous orchestra unfurled. The beginning of time, entrapped with the quantum idea of the real world, brought vulnerability and indeterminacy into the actual underpinnings of presence. Quantum particles moved in probabilistic circles, opposing the deterministic perfect timing imagined by traditional physical science. The slippery idea of time at the quantum scale indicated a more profound, more secretive association between the texture of the real world and the beginning of time.

As humankind wandered past the limits of Earth, the experience of time took on new aspects. The relativistic impacts of room travel, where the progression of time differed with the speed of the spectator, turned into a useful reality for space explorers venturing through the universe. The ticking of nuclear clocks on rocket uncovered the unpretentious yet quantifiable impacts of relativity, affirming that time itself was not a flat out steady however a dynamic and relevant peculiarity.

The investigation of far off universes, from the moon to the external scopes of the nearby planet group, divulged the land files of planetary time. Every cavity, each layer of silt, and each topographical element recounted an account of the enormous clock ticking ceaselessly through the ages. The beginning of time, engraved in the geographical record of divine bodies, turned into a period case for unwinding the secrets of the planetary group's set of experiences.

In the journey to comprehend a definitive starting points of the universe, the idea of vast time became the overwhelming focus. The Theory of prehistoric cosmic detonation, upheld by an abundance of observational proof, set that the universe had an unmistakable start — a second when reality rose up out of an early stage peculiarity. The beginning of time, in this astronomical beginning, denoted the commencement of the stupendous vast ensemble.

The early snapshots of the universe, the age of vast expansion, saw the quick development of spacetime at speeds outperforming the restrictions of light. The texture of the universe extended and unfurled, conveying with it the seeds of systems and the reverberations of the early stage fireball. The beginning of time, unpredictably associated with the unfurling show of enormous advancement, set up for the development of the designs we see in the huge astronomical display.

As the universe extended and cooled, matter blended into cosmic systems, stars, and planets. The dance of vast powers, administered by the laws of material science, molded the infinite embroidered artwork into the perplexing trap of designs we notice today. The grandiose microwave foundation, a weak shine pervading the universe, gave a depiction of the universe's initial minutes — a demonstration of the beginning of time itself.

In the immense enormous scope, universes shaped tremendous grandiose urban areas, each containing billions or even trillions of stars. Inside these heavenly cities, the dance of time proceeded. Stars were conceived, carried on with out their enormous lives, and in the long run met their destiny in the cauldron of heavenly advancement.

The beginning of time, engraved in the existence patterns of stars, turned into an enormous heartbeat resounding through the vast cityscape.

1.1 Introduction to Time as a Cosmic Concept

Time, an idea woven into the texture of presence, is a mystery that rises above the limits of human cognizance. It isn't simply a proportion of spans or a ticking clock; time is an infinite idea that penetrates the actual pith of the universe. From the greatness of cosmic developments to the tiny dance of particles, time coordinates the ensemble of presence on an inestimable scale.

In digging into the idea of time, one should explore the tremendous spread of the universe — an embroidery where worlds, stars, and planets waltz through the enormous expressive dance. The idea of time, as seen in ordinary human experience, is nevertheless a section of the more extensive enormous comprehension. To understand time as a grandiose idea, we should leave on an excursion that navigates the domains of physical science, reasoning, and the actual texture of spacetime itself.

The foundations of time as an infinite idea reach out to the beginning of the universe. As indicated by the predominant cosmological model, the Huge explosion, the universe started from a peculiarity — an imperceptibly little and thick place where the laws of material science separate. It was in this early stage pot that time itself was birthed. The grandiose clock, put into high gear right now of the Huge explosion, started ticking, coordinating the unfurling show of astronomical advancement.

As the universe extended, so did the extent of time. The early minutes, frequently covered in the grandiose obscurity of expansion, set up for the development of the designs we notice today. Systems, those grandiose islands holding onto billions of stars, arose as the aftereffect of gravitational collaborations and the recurring pattern of inestimable matter. Time, a fundamental member in this heavenly dance, directed the speed of astronomical development.

The enormous microwave foundation, a weak gleam pervading the universe, fills in as a remnant from the early ages of the universe. It is a preview of the universe when it progressed from a singing hot plasma to a state where particles could frame. The investigation of this old light not just gives bits of knowledge into the early states of the universe yet in addition fills in as a demonstration of the vast course of events — a timetable scratched by the hand of time itself.

In the vast cityscapes shaped by systems, time takes on a diverse job. Stars, the heavenly occupants of systems, are conceived, carry on with out their infinite lives, and in the long run meet their destiny. The existence patterns of stars, from the searing birth in heavenly nurseries to the dangerous passings in supernovae, are complicatedly attached to the progression of time. The astronomical heartbeat reverberations through the limitlessness of room, denoting the rhythms of heavenly advancement.

The dance of worlds, organized by the gravitational powers at play, is one more indication of time as an inestimable idea. Worlds impact and converge over infinite timescales, making new designs and modifying the astronomical scene. The development of cosmic designs more than billions of years lays out a picture of the vast timetable — a course of events where universes are the heroes in the terrific grandiose story.

Past the glowing domains of worlds lies the baffling area of dull matter. Albeit imperceptible to customary telescopes, the presence of dull matter is induced through its gravitational effect on apparent matter. The gravitational artful dance of dim matter, a grandiose dance arranged by the inconspicuous hand of a slippery substance, unfurls on timescales that inspire bigger thoughts. Dim matter, a vast puzzle by its own doing, adds one more layer to the multifaceted embroidery of time as an inestimable idea.

Wandering nearer to home, to the planetary domains inside systems, time shows itself in geographical files. Each layer of dregs, each geographical element, recounts an account of the inestimable clock ticking endlessly through the ages. The actual Earth, an observer to the ages, bears the scars and engravings of the progression of time. The investigation of Earth's topographical history gives a brief look into the more extensive vast story — a story written in the language of rocks and fossils.

As humankind investigates the universe, the experience of time takes on new aspects. The relativistic impacts of room travel, as anticipated by Einstein's hypothesis of relativity, become substantial real factors for space explorers.

The ticking of nuclear clocks on space apparatus uncovers the unobtrusive yet quantifiable impacts of relativity, affirming that time is definitely not a flat out steady yet a dynamic and logical peculiarity. The vast clock, impacted by the speed of the eyewitness and the gravitational fields experienced, turns into a sidekick on the excursion through the universe.

In the domain of quantum mechanics, the minute texture of reality reveals one more layer of the astronomical idea of time. Quantum particles, the structure blocks of issue, show ways of behaving that challenge traditional instincts. The probabilistic idea of quantum frameworks brings a component of vulnerability into the actual groundworks of presence. Time at the quantum scale turns into a subtle and secretive boundary, indicating a more profound association between the texture of the real world and the beginning of time itself.

The interaction among gravity and time, as clarified by the hypothesis of relativity, reshapes how we might interpret the grandiose idea of time. Gravity, the power that oversees the huge scope construction of the universe, impacts the progression of time. Gigantic articles twist the texture of spacetime, making gravitational wells that change the directions of divine bodies and the speed of time itself. The infinite dance of systems and the gravitational ensemble of the universe become indivisible from the elements of time.

The philosophical thoughts of old scholars, pondering the idea of time and its job in the human experience, track down reverberation in the cutting edge comprehension of time as an enormous idea. The bolt of time, the unidirectional stream from past to present to future, turns into a focal subject in the investigation of time's vast nature. The irreversible idea of time, combined with the inflexible walk towards entropy, shapes the infinite story into a story where time assumes a main part.

In the stupendous enormous ensemble, time isn't just a scenery; it is the director, coordinating the developments of heavenly bodies, the development of universes, and the dance of particles. An idea rises above the human-scale estimations of timekeepers and schedules. Time, as a grandiose idea, is woven into the actual texture of the universe — a string that ties the unique components of the universe into a lucid and dynamic entirety.

As we peer into the profundities of room and dig into the minuscule domains of quantum vulnerability, the grandiose idea of time uncovers itself as a bringing together guideline. An idea spans the perceptible and the minute, the noticeable and the imperceptible, the old and the contemporary. Time, in its enormous sign, welcomes us to ponder the idea of presence itself — an excursion that takes us past the limits of human discernment and into the grandiose profundities where the beginning of time unfurls in the entirety of its highness.

1.2 Theoretical Foundations: Einstein's Relativity and Quantum Mechanics

In the mid twentieth 100 years, the scene of hypothetical physical science went through a progressive change that reshaped how we might interpret the universe. Two mainstays of present day material science arose during this period — Albert Einstein's hypothesis of relativity and the quantum mechanics spearheaded by lights like Max Planck, Niels Bohr, and Werner Heisenberg. These hypothetical establishments laid the basis for another period in our cognizance of the central idea of room, time, and matter.

Einstein's hypothesis of relativity, created in two unmistakable plans — unique relativity and general relativity — tested old style thoughts of existence. Exceptional relativity, uncovered in 1905, presented the idea that the laws of physical science are no different for all spectators in unaccelerated movement. One of its earth shattering hypothesizes was that the speed of light in a vacuum is consistent for all onlookers, no matter what their movement comparative with the light source.

This apparently illogical thought prompted an outpouring of significant ramifications. The hypothesis delivered the idea of spacetime — a brought together four-layered structure wherein existence are indivisibly intertwined. Spacetime is definitely not a static stage yet a powerful field where the texture of reality can twist and curve affected by huge items.

The renowned condition

$\diamond$

=

$\diamond$

$\diamond$

2

E=mc

2

, determined by Einstein in his unique hypothesis of relativity, communicated the comparability among energy and mass. It uncovered that mass is a type of concentrated energy and made the way for the comprehension of atomic responses that power the sun and stars. Extraordinary relativity essentially adjusted our view of time, presenting the idea of time expansion — where time itself is capable contrastingly by onlookers in relative movement.

General relativity, introduced by Einstein in 1915, broadened the standards of unique relativity to envelop gravitational communications. In this hypothesis, gravity isn't a power sent across space, as imagined by Newton, yet a result of the distorting of spacetime brought about by mass and energy. Monstrous items, for example, stars and planets, direct the arch of the encompassing spacetime, affecting the ways of neighboring articles.

The prescient force of general relativity became apparent through its effective clarification of peculiarities like the twisting of light around huge articles, known as gravitational lensing, and the perihelion precession of Mercury's circle. The hypothesis likewise anticipated the presence of dark openings — locales where gravity is extraordinary to such an extent that nothing, not even light, can get away. These colorful expectations have been accordingly affirmed through cosmic perceptions.

While general relativity richly depicts the power of gravity for huge scopes, quantum mechanics arose as the hypothetical system administering the way of behaving of issue on the littlest scales. Quantum mechanics, started in the mid twentieth 100 years, was a takeoff from traditional physical science and presented a probabilistic and indeterministic perspective on the minute world.

The groundworks of quantum mechanics were laid by Max Planck, who presented the idea of quantization of energy to make sense of the range of blackbody radiation. He suggested that energy is quantized in discrete parcels, or "quanta," a takeoff from the traditional thought of constant energy.

Niels Bohr further created quantum hypothesis by proposing a model of the molecule that consolidated quantized electron circles. Bohr's model effectively made sense of the ghastly lines of hydrogen yet confronted difficulties in portraying more mind boggling molecules. The Copenhagen translation, planned by Bohr and Werner Heisenberg, presented the idea of wave-molecule duality, where particles like electrons show both wave and molecule qualities.

Heisenberg's vulnerability standard, a foundation of quantum mechanics, stated that specific sets of properties, like position and energy, can't be at the same time estimated with inconsistent accuracy. This characteristic vulnerability at the quantum level tested the deterministic perspective of old style physical science and highlighted the intrinsic probabilistic nature of quantum peculiarities.

The improvement of quantum mechanics went on with the detailing of Schrödinger's wave condition, which depicted the way of behaving of quantum frameworks as far as wave capabilities. The probabilistic translation of the wave capability, where the square of its sufficiency addresses the likelihood circulation of a molecule's situation, turned into a focal principle of quantum hypothesis.

Quantum mechanics presented the idea of superposition, where particles could exist in different states at the same time until estimated. This element led to the popular psychological study known as Schrödinger's feline, outlining the idiosyncrasies of quantum superposition.

Ensnarement, another particular quantum peculiarity, uncovered that particles could become connected so that the condition of one molecule momentarily impacts the condition of another, regardless of whether isolated by huge distances. This nonlocal association tested old style ideas of region and recommended a significant interconnectedness at the quantum level.

The advancement of quantum field hypothesis broadened the standards of quantum mechanics to incorporate the quantization of fields, regarding particles as excitations of these fields. The standard model of molecule material science, a victory of hypothetical physical science, effectively bound together electromagnetic, frail, and solid atomic powers inside the system of quantum field hypothesis.

Regardless of their singular victories, the speculations of relativity and quantum mechanics present a significant test while accommodating their standards. The tiny universe of quantum mechanics works in a domain where spacetime shape, a vital component of general relativity, is irrelevant. Endeavors to figure out a quantum hypothesis of gravity — a brought together structure that consolidates the standards of quantum mechanics and general relativity — have demonstrated subtle.

The quest for a hypothesis of everything, frequently alluded to as quantum gravity, stays an open outskirts in hypothetical material science. String hypothesis and circle quantum gravity are among the proposed ways to deal with accommodate the quantum and gravitational domains, each offering novel points of view on the principal idea of the universe.

String hypothesis sets that crucial substances are not point particles but rather minuscule, vibrating strings. The music of these vibrating strings compare to various molecule types, giving a brought together portrayal of every single known power and particles. String hypothesis presents additional aspects past the recognizable three spatial aspects and once aspect, recommending a secret intricacy to the texture of spacetime.

Circle quantum gravity, then again, approaches quantum gravity according to the viewpoint of discrete, quantized structures. In this structure, spacetime itself is granular, and the calculation of room rises up out of the communications of rudimentary units known as circles. Circle quantum gravity challenges the consistent idea of old style spacetime, offering an original point of view on the texture of the universe.

While these ways to deal with quantum gravity stay hypothetical and face difficulties in trial check, they address the bleeding edge of our journey to disentangle the most profound secrets of the universe. Hypothetical systems that bring together the standards of relativity and quantum mechanics hold the possibility to give a far reaching comprehension of the universe from the littlest sizes of quantum peculiarities to the immense scopes of inestimable designs.

The duality between the minuscule and perceptible domains, exemplified by the hypotheses of relativity and quantum mechanics, mirrors the complicated and interconnected nature of the universe. The enormous dance, coordinated by the standards of these speculations, winds around an embroidery where the texture of spacetime twists affected by huge items, and particles show ways of behaving that challenge old style instinct.

As how we might interpret the hypothetical groundworks of present day physical science keeps on advancing, so too does our view of the universe. The tradition of Einstein's relativity and the standards of quantum mechanics is engraved on our investigation of the universe, from the relativistic impacts experienced by space travelers to the quantum advancements molding the eventual fate of data handling.

The hypothetical underpinnings of current physical science not just act as scholarly systems for figuring out the universe yet in addition as guides for exploratory investigation. Molecule gas pedals, observatories, and space missions push the limits of our insight, testing the forecasts of relativity and examining the secrets of quantum peculiarities.

In the journey for a brought together hypothesis that embodies the standards of relativity and quantum mechanics, physicists explore the boondocks of the known and the unexplored world. The quest for quantum gravity, the investigation of dim matter and dim energy, and the mission to comprehend the starting points of the universe all stand as demonstration of the persevering through interest and scholarly ability of mainstream researchers.

As we stand on the cliff of disclosure, the hypothetical establishments laid by Einstein and the trailblazers of quantum mechanics remain signals enlightening the way ahead. The enormous idea of time, unpredictably interweaved with the standards of relativity and quantum mechanics, coaxes us to dive further into the grandiose secrets and open the insider facts that lie at the core of the universe.

1.3 Unraveling the Mysteries of the Big Bang

The universe, a sweeping and developing embroidered artwork, bears the permanent characteristic of its starting point — the Huge explosion. This particular occasion, covered in the fogs of grandiose relic, addresses the beginning of the universe as far as we might be concerned. The unwinding of the secrets encompassing the Enormous detonation is an excursion through the early stage fireball, a journey to grasp the vast birth that set up for the terrific show of grandiose development.

The idea of the Huge explosion started from the acknowledgment that the universe is extending. In the mid twentieth 100 years, the space expert Edwin Hubble, through perceptions of far off worlds, found a methodical connection between a system's distance and its redshift — a peculiarity where light from far off objects shifts toward longer frequencies.

Hubble's perceptions inferred that systems were getting away from one another, proposing a general extension of the universe.

This disclosure prompted the definition of the Theory of the universe's origin, an outlook changing idea that places the universe started from a very hot and thick state. As indicated by this hypothesis, the universe has been growing from that point onward, with worlds creating some distance from one another as the texture of space-time itself extends.

The mission to comprehend the earliest snapshots of the universe carries us to the age of inestimable expansion. Proposed by physicist Alan Guth in the mid 1980s, enormous expansion recommends that the universe went through an outstanding extension in the main parts of a second after the Huge explosion. This fast extending of spacetime settled a few riddles in cosmology, including the consistency of the enormous microwave foundation radiation and the huge scope design of the universe.

Inflationary cosmology sets that small quantum vacillations during the inflationary period were extended to grandiose scales, cultivating the arrangement of enormous designs like systems and world bunches. These quantum changes left an engraving on the vast microwave foundation, giving a window into the universe's early stage state.

The grandiose microwave foundation (CMB), found in 1965 by Arno Penzias and Robert Wilson, is a weak gleam penetrating the universe. It is the lingering heat from the early hot and thick period of the universe, presently cooled to only a couple of degrees above outright zero. The definite investigation of the CMB has turned into a foundation of current cosmology, offering a depiction of the universe when it was something like 380,000 years of age.

Satellites like the Enormous Foundation Voyager (COBE), the Wilkinson Microwave Anisotropy Test (WMAP), and the Planck satellite have planned the temperature varieties in the CMB with phenomenal accuracy. These temperature changes, engraved on the old light, give significant data about the sythesis, construction, and history of the early universe.

As we venture back in time, the universe turns into an undeniably hot and thick cauldron. At the unimaginably high temperatures of the early universe, particles and their antiparticles were constantly made and obliterated in a frenzy of high-energy cooperations. This period, known as the quark-gluon plasma age, went on for microseconds after the Huge explosion and addresses a stage where crucial particles existed as free substances.

The resulting change to the quark control age denoted a urgent change in the universe's development. As the universe extended and cooled, quarks and gluons, the structure blocks of protons and neutrons, joined to shape these baryonic particles. The universe, when a fuming ocean of rudimentary particles, progressed to a state overwhelmed by protons, neutrons, and different hadrons.

In the initial couple of moments after the Huge explosion, nucleosynthesis happened — a period when protons and neutrons joined to shape light components like hydrogen, helium, and little hints of lithium and beryllium. The exact wealth of these light components, as anticipated by nucleosynthesis models, gives a rigid trial of the Theory of prehistoric cosmic detonation. Perceptions of the early stage overflows adjust amazingly well to hypothetical forecasts, areas of strength for loaning to the Huge explosion structure.

The arrangement of iotas, explicitly hydrogen and helium, denoted one more basic point in grandiose advancement. At the point when the universe cooled to two or three thousand degrees, electrons could join with protons and neutrons to frame non-partisan iotas. This progress, known as recombination, prompted the decoupling of issue and radiation. Photons, at this point not dispersed by charged particles, started to travel unreservedly through space, leading to the straightforwardness of the universe to light.

The radiation discharged during recombination, known as the grandiose micro-wave foundation radiation, gives a vast fossil — a remnant from the early universe. Concentrating on the unpretentious temperature varieties in the CMB permits cosmologists to surmise the dissemination of issue in the early universe, uncovering the seeds that in the long run developed into the enormous designs we notice today.

The development of the primary stars and universes proclaimed the grandiose sun-rise — a period known as vast reionization. As the primary stars lighted, their extreme radiation ionized the encompassing hydrogen gas, changing it from an impartial to an ionized state. This age, happening around a billion years after the Enormous detona-tion, denoted a groundbreaking stage in the universe's set of experiences, changing it from an obscure to a straightforward state.

The infinite web, a huge organization of interconnected fibers and voids, arose as gravity brought matter into thick locales. Dim matter, an imperceptible and secretive substance that offsets noticeable matter in the universe, assumed a urgent part in this cycle. The gravitational draw of dim matter worked with the breakdown of gas mists, leading to the development of the primary worlds.

The investigation of far off systems and quasars, saw through strong telescopes like the Hubble Space Telescope, gives a brief look into the early universe. The identifica-tion of worlds at progressively higher redshifts permits cosmologists to test enormous ages nearer to the Huge explosion. The light from these far off objects conveys the en-graving of the mediating grandiose medium, uncovering the condition of the universe at various ages.

Worlds, as inestimable time cases, harbor data about the cycles forming the universe over now is the right time. The investigation of world advancement, from the early grandiose sunrise to the current day, discloses the complicated transaction between dim matter, customary matter, and the powers administering enormous designs. Huge scope studies, for example, the Sloan Computerized Sky Review, have planned large number of worlds, revealing insight into the astronomical web's tremendous and unpredictable design.

While the Theory of the universe's origin gives a strikingly fruitful structure to un-derstanding the perceptible universe's development, it likewise leaves specific inquiries unanswered. The idea of dull matter and dim energy, comprising around 95% of the universe, stays perhaps of the main secret in cosmology. The Theory of how things came to be, while making sense of the universe's development and the arrangement of light components, doesn't address the circumstances winning at the peculiarity — the mark of boundless thickness and temperature from which the universe arose.

The mission to comprehend the beginning of the universe and the circumstances close to the peculiarity requires a brought together hypothesis of quantum gravity — an undertaking that lies at the boondocks of hypothetical material science. String hypothesis, circle quantum gravity, and different methodologies try to accommodate the standards of quantum mechanics and general relativity, giving an extensive system to grasping the universe's introduction to the world and development.

Perceptions of the enormous microwave foundation, the huge scope construction of the universe, and the properties of systems keep on refining how we might interpret the Enormous detonation's fallout. The accuracy of current cosmological estimations, combined with the computational force of hypothetical models, empowers researchers to test the forecasts of the Theory of prehistoric cosmic detonation with extraordinary exactness.

In the vast journey for information, mechanical headways assume an essential part. Telescopes, both ground-based and space-based, furnished with modern instruments, permit stargazers to peer further into space and farther back in time. Molecule gas pedals and trials in underground research facilities add to how we might interpret crucial particles and their collaborations.

What's to come holds commitments of additional revelations and a more nitty gritty comprehension of the enormous beginning. The James Webb Space Telescope, set to send off before long, will open new boondocks in observational stargazing, giving bits of knowledge into the development of the primary cosmic systems and the airs of exoplanets. Headways in hypothetical physical science, pushed by progressing examination into quantum gravity and brought together speculations, offer the expectation of unwinding the secrets that continue past the astronomical skyline.

As we think about the excursion through the secrets of the Huge explosion, we end up at the juncture of perception, hypothesis, and mechanical development. The enormous account, written in the language of light, dim matter, and the concealed powers molding the universe, welcomes us to proceed with our investigation into the vast unexplored world. The unwinding of the Huge explosion's secrets, an adventure that started with the acknowledgment of the universe's development, is a continuous odyssey — a journey into the core of the universe, where the reverberations of the inestimable birth keep on resounding through the immense region of reality.

1.4 The Birth of Time in the Cosmos

In the tremendous breadth of the universe, where the dance of heavenly bodies unfurls against the setting of boundlessness, time arises as a major and baffling power. The introduction of time in the universe is an idea that rises above the ordinary ticking of clocks and the musical patterns of constantly — a grandiose ensemble organizes the unfurling show of presence itself. To unwind the secrets encompassing the introduction of time, one should leave on an excursion through the ages of inestimable development, from the early stage flames of the Huge explosion to the complicated dance of systems and the unpretentious rhythms of quantum reality.

The infinite story starts with the Enormous detonation, an occasion that denotes the origin of the universe as far as we might be concerned. The peculiarity, a place of limitless thickness and temperature, led to an inestimable blast that set the universe into movement. The universe, at first hot and thick, extended quickly, and as it did as such, it conveyed inside it the seeds of time. The actual texture of spacetime unfurled, making a phase whereupon the vast show would work out.

As the universe extended, it cooled, and the rudimentary particles framed in the early stage fireball. Quarks and gluons, the structure blocks of protons and neutrons, twirled in an enormous dance during the quark-gluon plasma age. This vaporous state, enduring just microseconds, denoted a stage where essential particles existed as free substances, and the universe was a fuming ocean of high-energy cooperations.

The change to the quark control age was an essential crossroads in grandiose history. As the universe extended and cooled further, quarks and gluons consolidated to shape protons, neutrons, and different hadrons. The universe, when a turbulent gathering of rudimentary particles, developed into a state overwhelmed by baryonic matter. This change, directed by the inflexible entry of time, established the groundwork for the resulting ages of vast advancement.

In the initial couple of moments after the Enormous detonation, nucleosynthesis happened — a cycle where protons and neutrons consolidated to frame light components like hydrogen, helium, and little hints of lithium and beryllium.

The exact overflows of these light components, as anticipated by nucleosynthesis models, offer an essential window into the circumstances winning in the early universe. The outcome of these expectations gives undeniable proof on the side of the Theory of the universe's origin.

The development of impartial molecules through recombination denoted one more achievement in the grandiose timetable. As the universe cooled to a couple thousand degrees, electrons joined with protons and neutrons, prompting the formation of nonpartisan hydrogen and helium. This progress permitted photons to travel uninhibitedly through space, starting the time of enormous straightforwardness. The grandiose microwave foundation radiation, a remnant from this time, fills in as a demonstration of the universe's initial minutes.

Recombination prepared for the development of the primary stars and systems — a period known as infinite sunrise. The extraordinary radiation from these early heavenly reference points started vast reionization, changing the early stage unbiased hydrogen into an ionized state. This age, happening around a billion years after the Huge explosion, addresses an essential stage in the universe's development, progressing it from an obscure to a straightforward state.

The gravitational dance of dull matter assumed a vital part in the development of enormous designs. Dim matter, a puzzling and imperceptible substance, offsets noticeable matter in the universe and gives the gravitational framework to the development of worlds, system groups, and the immense vast web. The inestimable designs that arose out of the interaction of dull matter and baryonic matter bear the engravings of the astronomical birth and the progression of time.

Observational devices, for example, strong telescopes and space tests, have permitted stargazers to look into the profundities of room and notice far off systems and inestimable peculiarities. The investigation of world development, from the inestimable sunrise to the current day, gives experiences into the mind boggling transaction between dim matter, customary matter, and the powers molding the enormous engineering. The grandiose web, an immense organization of interconnected fibers and voids, is a visual indication of the gravitational movement that has unfurled over infinite timescales.

The introduction of time in the universe is personally associated with the standards of Einstein's hypothesis of relativity. Exceptional relativity, revealed in 1905, presented a significant change in how we might interpret reality. The consistency of the speed of light and the idea of spacetime as a bound together system became foundations of this progressive hypothesis. General relativity, planned in 1915, stretched out these standards to portray gravity as the curve of spacetime brought about by mass and energy.

The exchange among gravity and time, as explained by broad relativity, acquaints a powerful viewpoint with the inestimable story. Enormous articles, for example, stars and worlds, twist the texture of spacetime, impacting the direction of neighboring items and adjusting the progression of time itself. The enormous dance of heavenly bodies, directed by the gravitational ensemble, turns into a sign of the complex connection between mass, energy, and the bend of spacetime.

In the domain of quantum mechanics, the introduction of time takes on an alternate aspect. Quantum hypothesis, created in the mid twentieth hundred years, depicts the way of behaving of issue and energy at the littlest scales. The vulnerability rule, wave-molecule duality, and the probabilistic idea of quantum frameworks present a degree of indeterminacy that challenges traditional thoughts of determinism.

The mission to comprehend the introduction of time at the quantum level prompts the investigation of spacetime's infinitesimal texture. String hypothesis, a hypothetical system that places basic substances as little, vibrating strings, tries to accommodate the standards of quantum mechanics and general relativity. String hypothesis presents additional aspects past the natural three spatial aspects and once aspect, offering a bound together portrayal of every single known power and particles.

Circle quantum gravity, one more way to deal with quantum gravity, treats spacetime as granular, with rudimentary units known as circles. This discrete construction challenges the consistent idea of old style spacetime, giving an extraordinary point of view on the texture of the universe. The introduction of time, as investigated through these quantum gravity speculations, turns into a significant investigation into the idea of reality at the littlest scales.

The vast birth of time, from the dangerous flames of the Enormous detonation to the complexities of quantum reality, welcomes consideration on the idea of presence itself. The introduction of time isn't simply an ordered unfurling of occasions; it is the actual substance of the inestimable orchestra that shapes the embroidery of the universe. Time, as an inestimable idea, rises above human estimations and turns into a necessary piece of the vast dance — a dance that started with the universe's unstable beginning.

As we think about the introduction of time in the universe, we wind up at the crossing point of logical request, philosophical consideration, and grandiose marvel. The continuous investigation of the infinite obscure, directed by the standards of relativity, quantum mechanics, and the boondocks of hypothetical material science, moves humankind into a more profound comprehension of the vast story.

Innovative progressions, from strong telescopes to molecule gas pedals, enable researchers to drive the limits of information and friend into the infinite profundities. Observational information, hypothetical models, and trial results meet to illustrate the introduction of time and the advancement of the universe.

The introduction of time in the universe is a demonstration of the inborn interest of the human soul. It entices us to wander further into the grandiose obscure, to unwind the excess secrets that cover the enormous beginning. The excursion through the introduction of time is an immortal odyssey — a journey that rises above the restrictions of human life and welcomes us to consider the glory of the universe and our place inside its tremendous and consistently developing embroidery.

Chapter 2

The Dance of Celestial Clocks

In the tremendous breadth of the universe, where reality entwine in a sensitive dance, heavenly timekeepers oversee the musicality of presence. The Dance of Heavenly Clocks is an enormous expressive dance, a multifaceted movement coordinated by the divine bodies that decorate the vast stage. Each heavenly element, from the brilliant stars to the confounding dark openings, assumes a part in this entrancing presentation.

At the core of this inestimable exhibition are the stars, those glowing guides that speck the velvet material of the night sky. Stars are the watches of the universe, their life cycles denoting the entry of ages. Brought into the world from the enormous nurseries of gas and residue, stars touch off in a blast of greatness, their atomic heaters producing components that will seed people in the future of heavenly bodies. As these vast behemoths exhaust their atomic fuel, they go through stupendous changes, coming full circle in touchy supernovae that dissipate their basic heritage across the universe.

The remainders of these divine firecrackers become the structure blocks for new stars, planets, and moons. The heavenly clock ticks on, and the cycle rehashes in a grandiose ensemble of birth, passing, and resurrection. Clouds, those ethereal billows of gas and residue, support the origin of stars, their whirling dance leading to the astonishing glow that lights up the universe. In the heavenly expressive dance, clouds are the stage whereupon stars are conceived, a demonstration of the everlasting pattern of creation.

Planets, the heavenly vagabonds that wander through the astronomical region, add their own rhythm to the Dance of Divine Timekeepers. In the boundlessness of room, planets follow curved circles around their parent stars, their gravitational dance forming the elements of whole planetary groups. Earth, a light blue speck suspended in the grandiose expressive dance, is observer to the multifaceted exchange of heavenly powers. The gravitational pulls and pulls from the sun and moon lead to the back and forth movement of tides, a cadenced dance that has endured starting from the earliest days of recorded history.

Moon, Earth's undaunted buddy, gives occasion to feel qualms about its brilliant shine the nighttime scene, its stages denoting the entry of lunar months. The Dance of Heavenly Timekeepers reaches out past our planetary group, embracing the bunch exoplanets that populate the cosmic system. Each exoplanet, with its one of a kind qualities and orbital peculiarities, contributes an unmistakable note to the vast song.

Past the recognizable domain of stars and planets, the Dance of Heavenly Timekeepers takes a more secretive turn with the incorporation of dark openings. These cryptic grandiose substances, brought into the world from the gravitational breakdown of monstrous stars, twist the texture of spacetime itself. Dark openings are the enormous maestros, directing the ensemble of gravity with an imperceptible mallet. They eat up issue and energy, making a gravitational bedlam that twists the actual embodiment of the real world.

The divine artful dance isn't without heavenly bodies challenge grouping. Pulsars, the remainders of cosmic explosion blasts, turn quickly, transmitting light emissions that range across the universe like beacon radiates. Their cadenced heartbeats add a throbbing thump to the grandiose dance, a heavenly metronome keeping time in the endlessness of room.

Systems, huge grandiose islands that harbor billions of stars, twisting through the universe in a great three step dance. The Dance of Heavenly Tickers reaches out to the inestimable scale, where systems connect and impact, molding the fate of the actual universe. Gravitational powers tie worlds into enormous bunches, making huge designs that stretch across incomprehensible distances.

Grandiose fibers, the imperceptible strings that weave the enormous embroidery, associate worlds in a huge vast web. The Dance of Heavenly Clocks is an excellent enormous show, unfurling on the phase of the universe with an accuracy that misrepresents its turbulent magnificence. Dull matter, the slippery substance that penetrates the universe, adds its baffling presence to the enormous scene, applying gravitational impact on the noticeable matter moves in its gravitational hug.

The Dance of Heavenly Timekeepers rises above the limits of our recognizable universe. In the domain of cosmology, the investigation of the huge scope design and development of the universe, researchers unwind the mysteries of the universe's introduction to the world and extreme destiny. The vast microwave foundation, the weak sparkle of radiation that saturates the universe, fills in as a grandiose fossil, offering hints about the universe's initial minutes.

The development of the universe, a basic part of inestimable advancement, moves cosmic systems from one another, extending the texture of spacetime itself. The Dance of Divine Tickers is entwined with the enormous heartbeat, the cadenced extension and constriction that characterize the universe's grandiose heartbeat. Dim energy, the puzzling power driving the sped up extension of the universe, adds an inestimable crescendo to the heavenly expressive dance.

The Dance of Divine Timekeepers isn't bound to the cool limitlessness of room. It stretches out its impact to the domain of time itself. Einstein's hypothesis of general relativity, a work of art of hypothetical material science, uncovers the cozy association among gravity and the shape of spacetime. In the astronomical expressive dance, gigantic divine bodies, similar to stars and dark openings, make gravitational wells that twist the texture of spacetime, impacting the progression of time itself.

Time expansion, an outcome of general relativity, implies that time elapses contrastingly in locales of differing gravitational force. The Dance of Heavenly Tickers presents a transient component, where time turns into a variable impacted by the enormous dance of gravity. Clocks on the outer layer of an enormous planet tick at an unexpected rate in comparison to those in the void of room, a relativistic three step dance that adds a worldly intricacy to the grandiose movement.

The Dance of Heavenly Tickers finds articulation in the astronomical cycles that administer the destiny of stars and systems. The grandiose dance of birth, life, and passing works out in the existence patterns of stars, from the blazing births in heavenly nurseries to the fantastic passings in cosmic explosion blasts. Universes, as well, go through inestimable transformation, blending and developing over vast timescales that bantam human appreciation.

As we think about the Dance of Divine Tickers, we are gone up against with the vast point of view of our own reality. The molecules that form our bodies were produced in the atomic heaters of long-dead stars, a demonstration of the interconnectedness of all grandiose matter. The Dance of Divine Timekeepers is certainly not a far off display however a necessary piece of our vast legacy.

In our mission to comprehend the Dance of Heavenly Timekeepers, humankind has sent off a huge number of telescopes and tests into the enormous unexplored world. Observatories in space, for example, the Hubble Space Telescope, have looked into the profundities of the universe, catching pictures of far off cosmic systems, nebulae, and other divine marvels. Mechanical adventurers, similar to the Mars meanderers, have wandered past Earth, revealing the mysteries of different planets in our planetary group.

The Dance of Heavenly Tickers has likewise allured human pilgrims to the edge of the universe. Space travelers, the cutting edge trailblazers of room, have circled the Earth, strolled on the moon, and wandered into the void of room. Their processes have extended how we might interpret the universe and our place inside it, overcoming any barrier between the earthbound and the extraterrestrial.

In the journey to disentangle the secrets of the grandiose dance, researchers have grown useful assets and instruments. Molecule gas pedals, for example, the Huge Hadron Collider, test the key structure blocks of issue, looking to figure out the idea of the universe at its generally principal level. Observatories furnished with state of the art innovation dive into the profundities of reality, catching the weak reverberations of the inestimable ensemble.

The Dance of Divine Tickers isn't just a logical undertaking yet in addition a well-spring of motivation for craftsmen, writers, and rationalists. The night sky, with its heavenly miracles, has enraptured the human creative mind for centuries. Old societies wove fantasies and legends around the heavenly bodies, finding in the stars the accounts of divine beings and legends. The Dance of Divine Clocks, as seen according to the point of view of Earth, has enlivened immortal masterpieces and writing, mirroring humankind's timeless interest with the universe.

In the twentieth hundred years, the approach of room investigation brought the Dance of Divine Tickers into the public cognizance. The famous pictures of Earthrise caught by the Apollo missions rose above public boundaries, offering a brief look at our common inestimable home. The light blue speck, as depicted via Carl Sagan, turned into an image of the delicacy and interconnectedness of life on The planet.

The Dance of Divine Timekeepers, with its enormous rhythms and heavenly harmonies, welcomes examination of humankind's position in the excellent embroidery of the universe. As we look at the night sky, we are stood up to with the boundlessness of room and the limitless

sizes of time. The Dance of Heavenly Tickers is an update that we are nevertheless transient spectators in the vast theater, observers to the unfurling show of the universe.

The investigation of the universe keeps on revealing new parts in the Dance of Divine Clocks. Progresses in innovation and hypothetical grasping open windows into beforehand unfamiliar domains. Gravitational wave observatories, like LIGO and Virgo, identify the waves in spacetime brought about by calamitous occasions like the crash of dark openings and neutron stars. These enormous quakes, reverberating through the texture of the universe, give another aspect to the infinite dance.

The Dance of Heavenly Tickers stretches out to the external compasses of the planetary group and then some. Explorer 1 and 2, mankind's farthest-arriving at tests, have wandered into the interstellar medium, conveying with them messages from Earth as the Brilliant Record. These messengers from humankind convey the tale of our reality, a demonstration of our interest and want to interface with the inestimable unexplored world.

In the quest for extraterrestrial life, the Dance of Heavenly Timekeepers takes on another importance. The disclosure of exoplanets inside the livable zone, where conditions for life as far as we might be concerned may exist, fills the creative mind with the chance of life past Earth. The inestimable expressive dance turns into a phase for the expected shows of outsider civilizations, each with its own infinite clock ticking to the musicality of its divine environmental elements.

The Dance of Heavenly Timekeepers likewise goes up against humankind with existential inquiries regarding the destiny of the actual universe. The extension of the universe, driven by dim energy, raises the ghost of a possible intensity demise, where the universe arrives at a condition of greatest entropy and every type of effort stops.

The grandiose dance, with its perplexing movement of creation and obliteration, might be bound for a last shade call.

However, even with the vast obscure, the Dance of Divine Timekeepers proceeds to enamor and motivate. The quest for information, the investigation of room, and the examination of the infinite embroidered artwork are attempts that rise above individual lifetimes. The vast dance, with its timeless rhythms, welcomes humankind to join the fantastic display of the universe and add to the unfurling story of the universe.

As we stand on the slope of another period of room investigation, with missions to Mars, the external planets, and then some, the Dance of Divine Clocks entices us to wander further into the inestimable unexplored world. The mission for figuring out, the hunger for information, and the human soul of investigation drive us to disentangle the secrets of the universe. The Dance of Heavenly Clocks, with its enormous rhythms and divine harmonies, welcomes us to be dynamic members in the continuous show of the universe.

All in all, the Dance of Heavenly Clocks is an enormous ensemble that works out across the tremendous region of the universe. From the introduction of stars in enormous nurseries to the destructive blasts of supernovae, from the cadenced beats of pulsars to the gravitational artful dance of dark openings, the divine dance is a demonstration of the magnificence and intricacy of the universe. The exchange of stars, planets, worlds, and the essential powers of the universe makes an entrancing display that rises above the limits of existence.

The Dance of Divine Clocks welcomes us to consider our spot in the great embroidered artwork of the universe. It provokes us to investigate the obscure, to look for replies to the essential inquiries that have interested humankind for a really long time. As we look at the night sky, we are helped that we are part to remember something a lot bigger, that the dance of the universe is a steadily unfurling story in which we assume an imperative part. The astronomical artful dance, with its timeless rhythms, allures us to join the dance and embrace the miracle of the universe.

2.1 Stellar Clocks: How Stars Mark the Passage of Time

In the huge heavenly region, stars stand as immortal sentinels, denoting the entry of ages with their iridescent presence. Heavenly Tickers, the divine watches of the universe, weave complex stories of birth, development, and downfall as they play out their vast artful dance. Stars, those brilliant spheres of atomic combination, hold the keys to interpreting the language of the universe. Their life cycles, throbs, and inevitable destiny make the parts out of a heavenly narrative that rises above the ages.

At the core of this heavenly story is the heavenly nursery, an inestimable cauldron where gas and residue combine to bring forth new stars. These interstellar nurseries, frequently tracked down inside goliath sub-atomic mists, are the supports of heavenly beginning. Gravitational powers shape these mists into thick centers, where the seeds of future stars lie lethargic, anticipating the enormous flash that will touch off their iridescent excursion.

As the sub-atomic mists breakdown under their own gravitational draw, the temperature and tension at their centers rise dramatically. A protostar arises, a heavenly baby encompassed by a case of gas and residue. The beginning star starts its gravitational pull, pulling in material from its environmental factors and making way for a heavenly presentation that will unfurl more than millions or even billions of years.

The Dance of Heavenly Tickers begins with the introduction of these protostars, as they go through the extraordinary course of gradual addition. The heavenly incipient organism develops, gathering mass and force, until the interior strain and temperature arrive at the place of atomic combination start. At this vital second, the protostar changes into a fundamental succession star, the stage in which it will spend most of its heavenly life.

Primary succession stars, similar to our own sun, are the workhorses of the universe, changing over hydrogen into helium through the course of atomic combination. This supported combination response produces the extreme intensity and light that characterize a star's splendor. The radiance and shade of a not set in stone by its mass, with bigger stars consuming more blazing and more splendid than their more modest partners.

The Dance of Divine Timekeepers uncovers itself in the ghostly marks of stars. Space experts, furnished with strong telescopes and spectrographs, analyze the starlight into an inestimable rainbow. These heavenly spectra divulge the compound organization, temperature, and radiance of stars, permitting stargazers to disentangle the mysteries encoded in the divine light. Each star, with its remarkable phantom unique mark, contributes a refrain to the stupendous divine sonnet.

As stars proceed with their principal succession presence, they keep a sensitive harmony between the gravitational power pulling internal and the tension produced by atomic combination pushing outward. This harmony characterizes the life span of a star in its fundamental grouping stage, a period that traverses millions to billions of years, contingent upon its mass. Our sun, a moderately aged star, has been sparkling consistently for around 4.6 billion years and is supposed to proceed with its iridescent presentation for another 5 billion years.

Be that as it may, the heavenly clock doesn't tick consistently for all stars. Huge stars, with more noteworthy gravitational powers and higher center temperatures, consume their atomic fuel at a sped up pace. These astronomical behemoths, however staggering in their splendor, have more limited primary grouping lifetimes. On the other hand, more modest stars, known as red smaller people, show an all the more comfortable speed, sparkling faintly yet persevering for trillions of years.

As the divine artful dance unfurls, stars in the end exhaust their essential fuel source in the center. For sun-like stars, this denotes the progress to the red goliath stage, a sensational change that proclaims the start of the end. The external layers of the star extend, immersing internal planets and adjusting the star's glow and temperature. The red goliath stage is a grandiose crescendo, a preface to the last venture in the existence of a sun-like star.

In the Dance of Divine Tickers, the red goliath stage fills in as an impactful sign of the fleetingness of heavenly presence. The external layers of the red monster are ousted into space in a wonderful showcase known as a planetary cloud, abandoning a little, thick center known as a white smaller person. The ousted material, enhanced with weighty components manufactured in the star's center, turns into the unrefined substance for people in the future of stars and planets.

For additional gigantic stars, the last sections of their heavenly adventure are set apart by unstable finales. These enormous titans go through cosmic explosion blasts, disastrous occasions that discharge energy identical to the brilliance of a whole universe briefly. Supernovae are the astronomical produces where components heavier than iron are integrated, dissipating these structure blocks across the universe.

The remainders of a cosmic explosion blast can appear in changed ways. At times, the center implodes under its own gravity, shaping a neutron star — a thick, city-sized object with gravity so extraordinary that a sugar-block estimated measure of neutron-star material would weigh as much as a mountain on The planet. In different occurrences, the center breakdown brings about the development of a dark opening, a vast chasm where gravity is solid to the point that nothing, not even light, can get away from its grip.

The Dance of Divine Clocks, notwithstanding, stretches out past the life and demise of individual stars. In the great embroidery of the universe, the remainders of heavenly blasts blend in with the early stage gas and residue of the interstellar medium, making a rich supply of material for people in the future of stars. This grandiose reusing process guarantees that the components fashioned in the red hot centers of long-dead stars track down their direction into the texture of new heavenly frameworks.

Heavenly bunches, gravitationally bound gatherings of stars brought into the world from a similar sub-atomic cloud, give a spellbinding group in the Dance of Divine Clocks. Open groups, with their free setups, exhibit stars of shifting ages and transformative stages. Globular groups, thickly stuffed and antiquated, harbor stars that have circled the cosmic community for billions of years. The investigation of heavenly groups disentangles the aggregate history of their constituent stars, offering a brief look into the unpredictable elements of heavenly networks.

The vast clock additionally appears in the musical throbs of specific stars. Variable stars, like Cepheids and RR Lyrae factors, go through normal changes in brilliance. These throbs are connected to interior cycles, permitting space experts to use them as enormous standard candles. By estimating the time of these throbs, stargazers can precisely decide the inborn brilliance of these stars, giving a solid technique to check distances across the universe.

The Dance of Heavenly Tickers stretches out its compass to paired star frameworks, where two stars circle a typical focus of mass. The gravitational transaction between these heavenly partners can prompt various peculiarities, including mass exchange, obscurations, and, surprisingly, the inevitable consolidation of the stars. Parallel frameworks offer experiences into the elements of heavenly associations, revealing insight into the complexities of divine organizations.

As we investigate the subtleties of heavenly timekeeping, the idea of a star's "metallicity" arises as a central member in the divine narrative. In astrophysical terms, metals incorporate all components heavier than helium. Stars with higher metallicity have more prominent overflows of these heavier components, showing a background marked by numerous ages of star development and cosmic explosion enhancement. The metallicity of a star impacts its qualities, including its tone, glow, and potential for facilitating planetary frameworks.

Heavenly Timekeepers, with their divine rhythms and grandiose harmonies, shape the actual texture of the universe. The components produced in heavenly centers, delivered into space through heavenly passings, become the unrefined substances for the development of planets, moons, and even life itself. The Dance of Divine Tickers entwines the fates of stars and the universes they occupy, making an orchestra of enormous extents.

In the mission to translate the language of the stars, stargazers utilize a bunch of observational devices and hypothetical structures. Telescopes, both ground-based and spaceborne, catch the light transmitted by stars across the electromagnetic range. Spectrographs analyze this light, uncovering the synthetic arrangement, temperature, and other significant boundaries of stars. Hypothetical models, controlled by computational reproductions, empower stargazers to disentangle the complex cycles happening inside heavenly insides.

Heavenly paleontology, a field that looks at the most established stars in the Smooth Manner, offers a one of a kind viewpoint on the early ages of the universe. Old stars, known as Populace II stars, have low metallicity, demonstrating that they shaped in a period before various ages of stars enhanced the universe with weighty components. Concentrating on these heavenly fossils gives a brief look into the circumstances that won in the universe's earliest stages.

The Dance of Heavenly Tickers additionally unfurls on the amazing size of universes. In the grandiose movement, universes, huge vast islands facilitating billions of stars, winding through the universe. The gravitational collaborations between cosmic systems shape their designs, leading to rich winding arms, curved shapes, and unpredictable structures. Systems, limited by gravity into groups and superclusters, make huge grandiose embroidered works of art that stretch across the inestimable void.

Supermassive dark openings, hiding in the hearts of worlds, contribute a grandiose bass note to the Dance of Divine Tickers. These gravitational behemoths, with masses millions or billions of times that of the sun, impact the elements of their host universes. As issue twistings into a dark opening's gravitational throat, strong planes of energy are sent off into space, chiseling the inestimable scene on a stupendous scale.

The enormous clock, administered by the persevering walk of time, likewise crosses with the more extensive area of cosmology. The investigation of the huge scope design and advancement of the universe uncovers the complicated interaction between dim matter, dull energy, and the apparent matter that comprises stars, systems, and enormous designs. The enormous microwave foundation, a weak sparkle penetrating the universe, fills in as a depiction of the universe's initial minutes, offering hints about its starting point and development.

In the Dance of Heavenly Timekeepers, the idea of grandiose expansion arises as a central member in the enormous story. Hypothetical models suggest that the universe went through a quick development in the principal minutes after the Huge explosion, streamlining inconsistencies and making way for the development of grandiose designs. The reverberations of this early stage extension resound in the grandiose microwave foundation, giving a window into the universe's early stages.

The Dance of Heavenly Timekeepers isn't restricted to the domains of perception and hypothesis. Human investigation, powered by interest and the longing to disentangle the secrets of the universe, broadens the span of the heavenly account. Space missions, both mechanical and monitored, adventure into the infinite obscure, catching amazing pictures of far off worlds, nebulae, and other divine miracles.

Space explorers, cutting edge pioneers, have circled the Earth, set foot on the moon, and wandered into the profundities of room. Their encounters give a special point of view on the enormous expressive dance, seeing the excellence of Earth from space and examining the immeasurability of the universe. The Dance of Divine Tickers, as seen from the vantage point of room, offers a groundbreaking perspective that rises above earthly limits.

In the continuous journey to investigate the vast wilderness, telescopes and observatories in space assume a crucial part. The Hubble Space Telescope, an innovative wonder circling over Earth's climate, has caught remarkable pictures of far off universes, nebulae, and other heavenly peculiarities. Space telescopes furnished with cutting edge instrumentation, for example, the James Webb Space Telescope, vow to uncover new parts in the heavenly annal.

The Dance of Heavenly Tickers likewise coaxes mankind to the external spans of the planetary group and then some. Automated tests, like Explorer 1 and 2, have left on interstellar excursions, conveying with them messages from Earth as the Brilliant Record. These messengers from humankind navigate the inestimable void, conveying the account of our reality and the perplexing rhythms of the heavenly dance to possible extraterrestrial audience members.

As we explore the enormous oceans, the Dance of Heavenly Tickers stands up to humankind with existential inquiries regarding the destiny of the actual universe. The speeding up extension driven by dull energy indicates an enormous predetermination where cosmic systems float separated, stars wear out, and the universe capitulates to a condition of most extreme entropy. The grandiose clock, ticking unavoidably, raises significant requests about a definitive course of the vast show.

However, notwithstanding these inestimable vulnerabilities, the Dance of Heavenly Timekeepers stays a getting through wellspring of motivation. It welcomes thought of our spot in the fabulous woven artwork of the universe and moves us to investigate the grandiose unexplored world. The quest for information, the mission for understanding, and the human soul of investigation drive us to unwind the secrets encoded in the heavenly rhythms.

In the enormous artful dance, the heavenly tickers act as signals, directing us through the divine scene. The Dance of Divine Clocks, with its timeless rhythms and enormous harmonies, is a demonstration of the excellence and intricacy of the universe. It allures us to join the excellent grandiose dance, to be dynamic members in the unfurling story of the universe.

As we stand near the very edge of another time of room investigation, with missions to far off planets, the external compasses of the planetary group, and then some, the Dance of Divine Clocks welcomes us to wander further into the enormous unexplored world. The heavenly rhythms, encoded in the illumination of far off stars, reverberation the heartbeat of the universe. The Dance of Heavenly Clocks calls upon humankind to investigate, find, and embrace the miracle of the universe, rising above the limits of existence.

2.2 Galactic Orbits and the Cosmic Calendar

In the huge grandiose scope, worlds rule as giant vast elements that shape the actual texture of the universe. Inside the complex dance of worlds, the idea of cosmic circles arises as a central component, impacting the infinite engineering on scales both terrific and minute. Cosmic Circles and the Enormous Schedule entwine, uncovering the unique interaction between divine bodies and the persevering entry of infinite time.

At the core of the inestimable expressive dance are cosmic systems, tremendous agglomerations of stars, gas, residue, and dim matter bound together by gravity. The Smooth Way, our home universe, is a banned twisting cosmic system with an intricate construction of winding arms, a focal bar, and a supermassive dark opening known as Sagittarius A* at its center. As worlds cross the vast stage, their constituent stars circle the cosmic focus in mind boggling designs, making heavenly harmonies that reverberation across the universe.

Cosmic circles, the complex pathways followed by stars inside a system, are represented by the gravitational powers applied by the cosmic mass dispersion. The enormous movement of cosmic circles relies upon the harmony between the centripetal power, pulling stars toward the cosmic focus, and the outward power created by the stars' orbital movement. These gravitational elements shape the rich twisting arms and focal lumps that portray winding worlds like the Smooth Way.

In the vast dance, stars display various orbital examples, impacted by elements like their separation from the cosmic focus and their underlying speeds. The cosmic plate, a leveled locale where most stars live, is a powerful field where stars follow curved, roundabout, or even tumultuous circles. The winding arms, portrayed by areas of improved star development, act as heavenly signs following the enormous pathways of stars.

The Dance of Divine Clocks stretches out to the grandiose size of universes, where whole frameworks circle each other in a heavenly three step dance. Universes, limited by gravity, structure system gatherings and bunches, making huge vast designs that stretch across colossal distances. The gravitational cooperations inside these vast combinations impact the directions of cosmic systems, encouraging a rich exchange that shapes the enormous scope design of the universe.

Cosmic superclusters, the biggest known structures in the universe, address the most fabulous stage in the Dance of Divine Clocks. These immense inestimable collections, comprising of various world bunches, characterize the vast web — an organization of interconnected fibers and voids that traverses the noticeable universe. The gravitational powers inside superclusters coordinate the developments of worlds on unfathomable scales, making a vast embroidery that mirrors the hidden construction of the universe.

In the investigation of cosmic circles, researchers dive into the vast schedule — a figurative system that relates the developmental phases of universes to the section of enormous time. The grandiose schedule gives a sequential point of view on the existence patterns of worlds, from their development to possible change or destruction. The ages set apart on this inestimable schedule unfurl as worlds advance, cooperate, and add to the powerful story of the universe.

Cosmic development, a focal topic in the grandiose schedule, envelops the different stages through which universes progress throughout enormous time. The existence pattern of a system includes cycles like star development, cosmic crashes, and the impact of supermassive dark openings. The Dance of Divine Clocks unfurls as universes navigate these stages, each stage leaving its engraving on the cosmic design and arrangement.

Twisting worlds, with their agile arms and brought together lumps, exhibit the richness of continuous star development. In the vast schedule, this stage addresses the dynamic young people of a world, where sub-atomic mists inside the winding arms bring forth new stars.

The heavenly nurseries inside these locales add to the enhancement of the system's synthetic creation, as progressive ages of stars manufacture heavier components through atomic combination.

As worlds age, their heavenly populaces develop, and the inestimable schedule changes to a more experienced stage. Curved worlds, described by a smooth and featureless appearance, frequently rise out of cosmic consolidations and cooperations. These infinite impacts, driven by gravitational powers, lead to the rebuilding of worlds and the arrangement of curved shapes. The Dance of Divine Clocks appears in the gravitational associations that shape the cosmic scene over tremendous astronomical timescales.

Supermassive dark openings, held onto inside the centers of numerous universes, assume a crucial part in the grandiose schedule. As issue twistings into these grandiose chasms, extreme radiation and strong planes are released, impacting the cosmic climate. Dynamic cosmic cores, energized by the gravitational growth of issue onto supermassive dark openings, address a unique stage in the grandiose schedule, molding the predetermination of their host worlds.

In the later phases of the vast schedule, universes might go through extinguishing, a cycle where star development reduces, and the cosmic populace ages. This calm stage denotes a change from the powerful energy of youth to a more repressed infinite presence. The Dance of Heavenly Tickers, be that as it may, doesn't stop. Universes keep on orbitting the enormous stage, their directions impacted by gravitational associations with adjoining vast buddies.

Cosmic crashes, an emotional episode in the grandiose schedule, add to the rebuilding of worlds and the development of new vast elements. At the point when universes consolidate, the Dance of Divine Clocks increases, making flowing powers that shape the morphology and arrangement of the connecting frameworks. These vast consolidations, noticed all through the universe, abandon morphologically unmistakable systems, frequently bearing the marks of their wild pasts.

In the fabulous organization of the astronomical schedule, systems are not lone elements but rather individuals from cosmic networks. System gatherings, comprising of a couple of universes limited by gravity, and world groups, facilitating various worlds inside a typical gravitational hug, act as inestimable get-togethers where cosmic communications are ordinary. The Dance of Divine Clocks inside these vast groups impacts the directions of worlds, encouraging a multifaceted interaction that shapes the grandiose scene.

Dim matter, a subtle and secretive part that offsets apparent matter in the universe, likewise takes part in the enormous dance. The gravitational impact of dim matter stretches out past the radiant areas of systems, chiseling the huge scope design of the astronomical web.

As systems circle inside this huge inestimable structure, they follow the undetectable gravitational strings woven by dim matter, adding intricacy to the multifaceted movement of cosmic circles.

The Dance of Heavenly Clocks unfurls in the noticeable domain of light as well as in the grandiose reverberations of gravitational waves. LIGO and Virgo, gravitational wave observatories intended to recognize the waves in spacetime brought about by calamitous occasions, offer another aspect to the investigation of cosmic elements. The consolidations of minimal articles, for example, neutron stars or dark openings, add to the enormous ensemble, making gravitational waves that navigate the universe.

The investigation of cosmic circles reaches out past the bounds of our Smooth Way. Exoplanetary frameworks, with their own perplexing elements of orbital movement, give bits of knowledge into the variety of astronomical models past our grandiose area. The disclosure of exoplanets, circling stars both inside and outside our universe, extends the vast account, indicating the potential for livable universes and the enormous overflow of planetary frameworks.

In the more extensive setting of cosmology, the Dance of Divine Clocks unfurls inside the enormous schedule of the actual universe. The idea of grandiose expansion, a hypothetical system depicting a quick extension in the early snapshots of the universe, shapes the enormous scope structure saw in the vast microwave foundation. The reverberations of this early stage extension, engraved on the infinite material, offer a worldly viewpoint on the inestimable schedule, connecting the universe's introduction to the world to its ensuing development.

The gravitational exchange between worlds, dull matter, and other grandiose parts keeps on molding the universe's predetermination. Dim energy, a puzzling power driving the sped up extension of the universe, adds a grandiose crescendo to the Dance of Heavenly Clocks. The destiny of the universe, affected by the persevering infinite schedule, suggests significant conversation starters about a definitive course of the universe's development.

As mankind adventures further into the vast obscure, the Dance of Divine Tickers welcomes consideration of our position in the fantastic woven artwork of the universe. The investigation of cosmic circles, gravitational waves, and the infinite schedule upgrades how we might interpret the unique cycles molding the universe. Observatories, both ground-based and spaceborne, catch the inestimable light and gravitational waves that convey the narratives of universes across enormous ages.

Chasing information, the human soul of investigation energizes the mission to disentangle the secrets encoded in the Dance of Divine Timekeepers. Mechanical tests, telescopes, and future space missions vow to uncover new sections in the enormous account, revealing insight into the complexities of cosmic circles and the astronomical schedule.

The exchange between gravity, dim matter, and noticeable matter in the excellent heavenly expressive dance keeps on spellbinding our aggregate creative mind and push mankind toward a more profound comprehension of the universe's complicated dance.

2.3 Black Holes as Temporal Distortions

In the enormous embroidered artwork of the universe, dark openings stand as confounding divine articles that challenge how we might interpret space, time, and gravity. Past their puzzling occasion skylines lies a domain where the normal guidelines of material science separate, and the texture of spacetime itself is distorted into incomprehensible setups. Dark openings, with their extreme gravitational fields, act as transient twists — distorting the actual substance of time and guiding us into an enormous domain where the customary progression of fleeting flows is disturbed.

At the core of a dark opening dwells a peculiarity, a place of limitless thickness where the known laws of material science quit giving significant depictions. This peculiarity is covered by the occasion skyline, a limit past which departure becomes unthinkable because of the mind-boggling gravitational draw. The occasion skyline denotes the final turning point, catching everything, including light, inside its gravitational grasps.

The Dance of Heavenly Timekeepers takes on a remarkable mood nearby dark openings. As an item moves toward the occasion skyline, time itself seems to dial back according to the point of view of a far off eyewitness. This impact, known as gravitational time enlargement, emerges from the twisting of spacetime brought about by the extraordinary gravitational field close to the dark opening. The nearer an item is to the occasion skyline, the more articulated the easing back of time becomes.

Albert Einstein's hypothesis of general relativity, the primary structure for figuring out gravity, predicts gravitational time widening as a result of huge items twisting the texture of spacetime. Within the sight of a dark opening, the bend of spacetime turns out to be outrageous to such an extent that opportunity seems to arrive to a virtual halt close to the occasion skyline. This peculiarity isn't simply hypothetical; it has been tentatively affirmed through perceptions of gravitational time enlargement close to huge heavenly bodies.

For an onlooker arranged a long way from the gravitational impact of a dark opening, time seems to advance at its not unexpected rate. Notwithstanding, a spectator falling into a dark opening would encounter an alternate reality. As they approach the occasion skyline, the gravitational time expansion impact turns out to be progressively articulated. According to the viewpoint of the falling eyewitness, the outer universe goes through quick changes, while their own insight of time remains moderately unaltered. This relativistic impact indicates the significant ramifications dark openings hold for how we might interpret time as an infinite consistent.

The Dance of Heavenly Tickers takes on a much really fascinating beat while considering the peculiarity of time widening close pivoting dark openings. As indicated by the Kerr arrangement of Einstein's situations, dark openings can have precise force, prompting their pivot. Thus, these turning dark openings show an unexpected impact known as edge hauling or the Lense-Thirring impact. This gravitational impact causes spacetime to twirl around the turning dark opening, hauling along neighboring items.

Nearby a pivoting dark opening, the Lense-Thirring impact further twists the texture of spacetime, impacting the circles of encompassing matter. The pivot of the dark opening bestows a contort to the encompassing spacetime, making a peculiarity likened to an infinite whirlpool. This rotational angle adds one more layer to the transient twists experienced close to dark openings, acquainting new intricacies with the Dance of Divine Timekeepers.

The outrageous circumstances close to dark openings lead to an interesting component known as gravitational redshift. This impact, anticipated by broad relativity, happens when light produced from a source close to a huge item, for example, a dark opening, encounters a shift toward longer frequencies as it moves out of the gravitational well. Gravitational redshift not just gives a device to space experts to test the gravitational conditions around dark openings yet in addition adds to the worldly bends related with these grandiose substances.

As an article falls toward a dark opening, the rising gravitational draw causes the light it discharges to encounter gravitational redshift. For an outer eyewitness, the light from the falling article shows up progressively redshifted as it moves toward the occasion skyline. As time enlargement and gravitational redshift entwine close to dark openings, the Dance of Heavenly Clocks appears as a hypnotizing show of relativistic impacts, changing the impression of time in the vast area of these confounding items.

One of the most notable expectations of general relativity connected with dark openings is the peculiarity of gravitational time widening making time actually stop at the occasion skyline. This interesting idea is known as "time freezing." As an item moves toward the occasion skyline, the gravitational time expansion turns out to be outrageous to such an extent that time seems to dial back limitlessly according to the viewpoint of a far off spectator. In this situation, an outer spectator would observer a clock close to the occasion skyline ticking increasingly slow until it appears to freeze through and through.

The idea of time freezing close to a dark opening brings up significant issues about the idea of time itself. It challenges our instinctive comprehension of time as a wide-spread consistent, recommending that under outrageous gravitational circumstances, time can turn into a pliable and relative idea. The Dance of Heavenly Clocks, as seen close to dark openings, welcomes us to mull over the actual texture of spacetime and the essential idea of fleeting reality.

One of the interesting parts of dark openings as transient contortions is the chance of time travel. While the idea of time travel has for quite some time been a staple of sci-fi, the hypothetical system of general relativity considers situations where time widening close to a dark opening could prompt time travel-like impacts. These situations include articles or spectators encountering various paces of time contingent upon their vicinity to the dark opening.

One such speculative situation includes an infinite peculiarity known as a safe wormhole. Wormholes are speculative passages in spacetime that might actually associate far off locales of the universe. In the event that a navigable wormhole existed close to a dark opening, an item or spectator entering the area of the dark opening could encounter huge time widening. After rising up out of the district impacted by the dark opening, they could end up in an alternate transient reference outline, really encountering time travel into what was in store contrasted with onlookers who stayed external the impact of the dark opening.

In any case, the idea of time travel close to dark openings accompanies its portion of intricacies and conundrums, generally quite the well known "granddad mystery." That's what this oddity proposes assuming that time travel were conceivable, an explorer might actually change occasions previously, prompting inconsistencies and consistent irregularities. The goal of such mysteries lies at the convergence of hypo-thetical material science, quantum mechanics, and our advancing comprehension of the basic idea of time.

The Dance of Heavenly Tickers, as affected by dark openings, stretches out its impact to the more extensive inestimable scene. The gravitational waves transmitted by the cooperations of gigantic items, like blending dark openings, give a one of a kind instrument to examining the idea of spacetime. Gravitational wave observato-ries, including LIGO and Virgo, identify the waves in spacetime brought about by destructive occasions, offering another window into the fleeting twists related with dark openings.

The discovery of gravitational waves from dark opening consolidations gives not just an affirmation of Einstein's hypothesis of general relativity yet additionally an immediate perception of the fleeting waves made by the grandiose dance of these mysterious substances. The gravitational waves produced by dark opening consolidations spread through the texture of spacetime, conveying with them data about the majority, turns, and different properties of the elaborate dark openings. The exact estimation of these gravitational waves opens new roads for understanding the fleeting bends related with dark openings.

As we dive further into the vast obscure, the investigation of dark openings and their fleeting contortions keeps on pushing the limits of how we might interpret the universe. The investigation of supermassive dark openings at the focuses of cosmic systems, incorporating the one dwelling in our own Smooth Manner, gives experiences into the complicated communications between dark openings and their general surroundings.

The Dance of Divine Tickers stretches out its compass to the investigation of cosmic cores, where supermassive dark openings impact the elements of stars circling in their gravitational hug.

The observational procedures utilized to concentrate on dark openings include a great many electromagnetic frequencies, from radio waves to X-beams. Telescopes and observatories, both ground-based and spaceborne, catch the light and different signs transmitted by issue as it twistings into dark openings. These perceptions permit researchers to interpret the interesting marks of dark opening connections and disentangle the fleeting bends engraved on the inestimable material.

The continuous journey to comprehend dark openings as transient contortions likewise includes hypothetical undertakings. The advancement of numerical models and reenactments empowers researchers to investigate the complicated elements of issue in the gravitational field of dark openings. Mathematical reproductions give a virtual research center to concentrating on the fleeting contortions related with dark openings, permitting researchers to reenact the perplexing transaction between gravity, matter, and spacetime ebb and flow.

In the excellent enormous artful dance, the investigation of dark openings as worldly mutilations meets with the more extensive scene of hypothetical material science. The mission for a brought together hypothesis that accommodates general relativity with quantum mechanics — the two mainstays of current physical science — vows to reveal insight into the idea of dark openings at the quantum level. The goal of the supposed "data oddity," which concerns the destiny of data falling into a dark opening, stays a boondocks in how we might interpret the enormous dance of these puzzling substances.

The Dance of Divine Tickers, as impacted by dark openings, stretches out its impact to the philosophical domain. The significant ramifications of time enlargement, fleeting twists, and the potential for time travel challenge our impression of the real world and the idea of presence. The vast dance close to dark openings entices us to ponder the actual substance of time, welcoming a combination of logical request and magical investigation.

As we explore the inestimable oceans, the Dance of Heavenly Timekeepers defies mankind with the many-sided interchange between dark openings and the texture of spacetime. Space missions and observatories, both existing and arranged, vow to disclose new sections in the vast story, catching the quintessence of dark openings and their worldly bends. The investigation of dark openings welcomes us to leave on an excursion to the core of enormous secrets, where time itself turns into a variable in the terrific grandiose dance.

2.4 Gravitational Waves: Ripples in the Fabric of Time

In the huge vast field, where heavenly bodies get into the rhythm of gravity, a grandiose orchestra unfurls. Inside this complex dance, a momentous peculiarity becomes the dominant focal point: gravitational waves. These subtle waves in the texture of spacetime are delivered by probably the most destructive occasions known to man, conveying with them data about the elements of enormous items and the idea of gravity itself. Gravitational waves, the couriers of grandiose connections, open another window into the secret domains of the universe, uncovering the rich woven artwork of spacetime in its dynamic magnificence.

First anticipated by Albert Einstein in 1916 as an outcome of his hypothesis of general relativity, gravitational waves are aggravations in the curve of spacetime brought about by the speed increase of enormous items. As indicated by Einstein's hypothesis, gravity isn't a power communicated through space, as imagined by Isaac Newton, yet rather a distorting of spacetime itself. Monstrous articles, like planets, stars, and dark openings, make "imprints" in this spacetime texture, and when these items move or collaborate, they produce swells — gravitational waves — that spread through the universe.

The Dance of Divine Clocks, as impacted by gravitational waves, adds another layer to the grandiose orchestra. Gravitational waves give an interesting method for concentrating on the universe, offering experiences into occasions and peculiarities that are generally stowed away from customary adaptive perceptions. These waves in spacetime travel at the speed of light, conveying data about the sources that delivered them and the idea of the spacetime through which they proliferate.

One of the most sensational wellsprings of gravitational waves is the crash and consolidation of reduced objects, for example, dark openings and neutron stars. At the point when these huge bodies circle one another and ultimately blend, they emanate strong eruptions of gravitational waves that echo through the universe. The identification of these waves opens another time in observational stargazing, permitting researchers to "tune in" to the astronomical soundscape made by these brutal vast occasions.

The time of gravitational wave space science formally started with the weighty discovery of gravitational waves by the Laser Interferometer Gravitational-Wave Observatory (LIGO) on September 14, 2015. LIGO, a couple of indistinguishable finders — one in Livingston, Louisiana, and the other in Hanford, Washington — was intended to quantify the little extending and pressing of spacetime brought about by passing gravitational waves. The noteworthy occasion denoted the principal direct perception of gravitational waves, affirming a critical expectation of Einstein's hypothesis of general relativity.

The debut location by LIGO was the consequence of the impact and consolidation of two dark openings, each multiple times the mass of the sun. As these dark openings spiraled toward one another, they delivered energy as gravitational waves. In the last part of a second prior to consolidating, the gravitational waves arrived at a force that could be identified by LIGO, creating an unmistakable tweet like sign. The perception affirmed the presence of gravitational waves as well as given the principal direct proof of parallel dark opening consolidations.

The Dance of Heavenly Timekeepers, as uncovered by gravitational waves, stretches out its compass to the investigation of neutron stars — thick remainders of huge stars that have depleted their atomic fuel. Neutron star consolidations, one more intense wellspring of gravitational waves, offer an interesting an open door to investigate outrageous conditions of issue and unwind the secrets of the atomic physical science that oversee these vast items. The impact and consolidation of neutron stars additionally hold the way to figuring out the starting points of weighty components, like gold and platinum, in the universe.

A milestone second in gravitational wave stargazing happened on August 17, 2017, when LIGO and its European partner, Virgo, distinguished a gravitational wave signal starting from the consolidation of two neutron stars. In contrast to the past dark opening consolidations, this occasion, known as GW170817, was joined by an explosion of electromagnetic radiation, including gamma-beam explodes and optical signs. The multimessenger perception, including both gravitational waves and customary adjustable perceptions, denoted a critical headway in how we might interpret grandiose occasions.

The Dance of Heavenly Timekeepers likewise includes the investigation of paired frameworks where one of the parts is a pulsar — an exceptionally charged, pivoting neutron star. Pulsars produce light emissions radiation that are seen as standard heartbeats as the star turns. As these pulsar-neutron star doubles circle one another, they produce gravitational waves, slowly losing energy and making the orbital period decline. The investigation of these twofold frameworks gives a special research facility to testing the expectations of general relativity and refining how we might interpret the properties of neutron stars.

Gravitational waves, with their capacity to cross tremendous inestimable distances unrestricted, convey with them the possibility to divulge the mysteries of the early universe. The infinite microwave foundation (CMB), a weak sparkle that pervades the universe, addresses a preview of the universe's state around 380,000 years after the Enormous detonation. Gravitational waves produced during the inflationary age — a quick extension of the universe in its initial minutes — leave particular engravings on the polarization examples of the CMB. The location of these early stage gravitational waves could give experiences into the essential cycles that molded the universe in its outset.

In the journey to investigate the grandiose wildernesses, gravitational wave observatories are persistently pushing the limits of responsiveness and accuracy. High level LIGO and High level Virgo, moves up to the underlying locators, have fundamentally worked on their capacity to identify gravitational waves. These progressions, combined with plans for future observatories, for example, the Laser Interferometer Space Radio wire (LISA), vow to open new parts in our investigation of the gravitational wave universe.

LISA, a space-based gravitational wave observatory arranged by the European Space Organization (ESA), plans to distinguish gravitational waves in the low-recurrence system, supplementing the capacities of ground-based finders like LIGO and Virgo. LISA will comprise of three shuttle flying in a three-sided development, each lodging quickly dropping test masses. Gravitational waves going through spacetime will cause minute changes in the distances between these test masses, permitting LISA to distinguish an expansive scope of gravitational wave signals.

The Dance of Heavenly Timekeepers unfurls across a range of gravitational wave frequencies, each noteworthy various features of the vast story. Ground-based identifiers like LIGO and Virgo are delicate to higher-recurrence waves delivered by smaller article consolidations, while space-based observatories like LISA are intended to catch lower-recurrence waves created by gigantic infinite occasions, for example, the inspiral and consolidation of supermassive dark openings at the focuses of systems.

The investigation of gravitational waves as couriers from the universe has extended how we might interpret divine peculiarities as well as given another means to test the central precepts of material science. Gravitational wave perceptions offer rigid trial of general relativity, testing the idea of gravity in the solid field system. The arrangement between the noticed signs and the forecasts of general relativity has, up to this point, reaffirmed the power of Einstein's hypothesis in depicting the gravitational collaborations of enormous articles.

The Dance of Heavenly Timekeepers, as seen through gravitational waves, additionally holds the commitment of revealing new and intriguing enormous peculiarities. The recognition of middle of the road mass dark openings, whose presence lies between heavenly mass dark openings and supermassive dark openings, stays a tempting chance. Gravitational wave signals from the inspiral and consolidation of these halfway mass dark openings could give important experiences into the slippery populace of dark openings that occupy the infinite scene.

The investigation of twofold frameworks including white midgets — degenerate leftovers of low-mass stars — addresses one more road for gravitational wave stargazing.

These frameworks, known as twofold white smaller people, discharge gravitational waves as they circle one another, in the long run spiraling toward consolidation.

The location of gravitational waves from twofold white bantam consolidations improves how we might interpret heavenly development as well as offers a likely begetter for a class of heavenly blasts known as Type Ia supernovae.

The Dance of Heavenly Clocks, as enlightened by gravitational waves, stretches out its impact to the investigation of supermassive dark openings dwelling at the focuses of cosmic systems. These behemoths, with millions or billions of times the mass of the sun, impact the elements of their host worlds. At the point when two supermassive dark openings circle one another and in the long run blend, they produce gravitational waves that convey remarkable marks. The identification of such occasions gives critical bits of knowledge into the coevolution of systems and their focal dark openings.

In the investigation of vast peculiarities through gravitational waves, researchers are additionally wandering into the domain of multimessenger space science. The organized perception of heavenly occasions through both gravitational waves and conventional electromagnetic channels, like noticeable light and radio waves, offers a more exhaustive comprehension of the fundamental astrophysical cycles. Multimessenger perceptions improve the accuracy of astrophysical estimations and empower the investigation of vast occasions across a more extensive range of frequencies.

The Dance of Heavenly Timekeepers, as seen through gravitational waves, keeps on dazzling the creative mind of researchers and the public the same. The recognition of these subtle waves in spacetime has introduced another period of disclosure, uncovering enormous peculiarities that were recently stowed away from our observational reach. Gravitational wave cosmology has turned into an incredible asset for testing the principal parts of the universe and unwinding the secrets encoded in the enormous dance of huge items.

As we explore the vast oceans, the Dance of Divine Timekeepers allures humankind to investigate the unfamiliar domains of the gravitational wave universe. Space missions, observatories, and mechanical headways vow to additionally extend our capacity to pay attention to the vast ensemble made by the communications out of huge items. The quest for information, driven by the longing to grasp the idea of gravity and the privileged insights of the universe, moves us into a future where gravitational waves keep on uncovering the secret rhythms of the divine dance.

Chapter 3

Biological Rhythms and Earth's Timekeepers

Natural rhythms, characteristic worldly examples that oversee different physiological and social cycles in living organic entities, are unpredictably entwined with Earth's timekeeping components. These rhythms, frequently alluded to as circadian rhythms, direct a plenty of capabilities, going from rest wake cycles to chemical emission and metabolic exercises. Understanding the significant association between natural rhythms and Earth's watches is urgent for grasping the sensitive equilibrium that supports life on our planet.

At the center of natural rhythms is the circadian clock, an inward timekeeping framework that synchronizes a living being's exercises with the 24-hour day-night cycle. The expression "circadian" is gotten from the Latin words "around" (signifying 'around') and "diem" (signifying 'day'), underscoring the roughly 24-hour nature of these rhythms. While the circadian clock is endogenous, meaning it starts inside a life form, it is significantly affected by outer signals, with the most intense one being the light-dim cycle.

The World's turn on its pivot makes an anticipated pattern of light and obscurity, bringing about a 24-hour day. This ecological prompt fills in as an essential synchronizer for circadian rhythms, a peculiarity known as entrainment. The expert circadian clock in vertebrates, including people, is situated in the suprachiasmatic core (SCN) of the nerve center. The SCN gets input from the eyes' photoreceptor cells, especially delicate to light, permitting it to adjust the interior circadian mood to the outside light-dull cycle.

The complex connection between organic rhythms and Earth's watches stretches out past the circadian clock. Occasional changes, driven by the World's hub slant and circle around the sun, present extra fleeting varieties that impact the way of behaving and physiology of living life forms. These more drawn out term rhythms are vital for cycles like proliferation, relocation, and hibernation.

One wonderful illustration of the association between organic rhythms and occasional changes is tracked down in the conceptive procedures of numerous species. Because of varieties in day length, creatures change their conceptive exercises to boost their posterity's possibilities of endurance. This peculiarity, known as photoperiodism, is exemplified by the rearing times of various species, including birds, well evolved creatures, and bugs. The exact timing of generation guarantees that the youthful are conceived or incubated when natural circumstances, like food accessibility and temperature, are ideal.

Relocation designs likewise mirror the impact of Earth's watches on organic rhythms. Numerous species, from birds to whales, attempt occasional movements to take advantage of positive natural circumstances. The planning of these excursions is frequently finely tuned to the evolving seasons, guaranteeing that transient courses line up with the accessibility of assets and reasonable settling or reproducing locales.

Hibernation, a condition of lethargy described by a critical decrease in metabolic action and energy consumption, is one more illustration of the mind boggling exchange between natural rhythms and Earth's watches. Sleeping species enter this state to save energy during times of natural affliction, like winter when food is scant. The commencement and end of hibernation are arranged by inward organic timekeepers that answer signals like temperature and day length.

Notwithstanding the circadian and occasional rhythms, lunar rhythms likewise assume a part in forming natural examples. The gravitational impact of the moon, combined with its evolving stages, has been recommended to influence specific ways of behaving in marine life forms, a peculiarity known as the lunar or flowing cadence. For instance, numerous types of marine life show synchronized conceptive exercises and movements connected to the lunar cycle.

The entwining of organic rhythms with Earth's watches isn't bound to the collective of animals. Plants additionally show circadian rhythms, impacting cycles like photosynthesis, supplement take-up, and blooming.

The planning of blossoming, specifically, is a basic part of a vegetation's cycle, and it is frequently organized with natural variables, including day length and temperature.

The sub-atomic systems fundamental circadian rhythms have been a subject of broad examination. At the core of these systems are hereditary criticism circles including a bunch of "clock qualities." These qualities produce proteins that collaborate in a complicated way, shaping an administrative organization that creates the cadenced articulation of different qualities over the course of the day. The revelation of these atomic parts has given important experiences into how organic entities keep inward transient control and synchronize with outer ecological prompts.

The complicated dance of clock qualities includes positive and negative criticism circles. In the positive circle, clock qualities enact the record of different qualities, including those that code for clock proteins. As these proteins collect, they, thusly, restrain the action of the clock qualities, making a negative criticism circle. The repeating rise and fall of these sub-atomic players bring about the cadenced articulation of qualities associated with various physiological cycles, arranging the recurring pattern of an organic entity's day to day exercises.

While the sub-atomic premise of circadian rhythms is monitored across a great many animal categories, there is eminent variety in the particular parts and administrative pathways. This variety features the versatility of organic tickers to various natural specialties and environmental systems. For instance, a few creatures might have varieties in the length of their circadian beat, permitting them to more readily line up with explicit ecological circumstances.

The impact of natural rhythms stretches out to human wellbeing and prosperity. Interruptions to circadian rhythms, for example, those brought about by shift work, fly slack, or sporadic rest designs, can significantly affect physiological capabilities. The circadian misalignment, where inward natural tickers are in conflict with outer prompts, has been related with different medical problems, including rest issues, metabolic aggravations, and an expanded gamble of persistent illnesses.

In present day culture, where fake lighting and nonstop exercises have turned into the standard, there is a developing acknowledgment of the significance of keeping a sound circadian mood. Methodologies, for example, openness to normal light during the day, laying out ordinary rest wake cycles, and limiting light openness around evening time are prescribed to help circadian wellbeing. The area of chronobiology, committed to the investigation of organic rhythms, keeps on uncovering the complexities of these inside timekeeping instruments and their suggestions for human wellbeing.

The job of organic rhythms in psychological well-being is an arising area of examination. Circadian disturbances have been connected to temperament problems, like wretchedness and bipolar issue.

The circadian framework collaborates with synapses, chemicals, and cerebrum areas ensnared in mind-set guideline. Understanding these associations might make ready for novel helpful intercessions for psychological well-being conditions.

The complicated connection between organic rhythms and Earth's watches additionally has suggestions for the field of chronotherapy, where the planning of clinical mediations is improved in view of circadian standards. The viability of specific meds, as well as the body's reaction to medicines, can differ contingent upon the hour of day. Fitting clinical intercessions to a person's circadian beat holds the possibility to upgrade treatment results and limit incidental effects.

The investigation of organic rhythms has broad ramifications past the limits of Earth. As humankind wanders into space investigation, understanding what the shortfall of Earth's regular watches means for natural rhythms becomes central. Space missions, with their microgravity surroundings and absence of a predictable day-night cycle, present difficulties to the support of circadian rhythms among space travelers. Research in this space looks to foster procedures to moderate the effect of room travel on the wellbeing and prosperity of spacefarers.

The intriguing interaction between organic rhythms and Earth's watches additionally reaches out to the domain of environment and protection. Natural changes, including environmental change and living space disturbance, can modify the signs that organic entities depend on for entrainment. These progressions might prompt jumbles between natural rhythms and ecological circumstances, influencing vital life cycles like movement, propagation, and taking care of ways of behaving.

The investigation of natural rhythms is innately interdisciplinary, drawing on bits of knowledge from hereditary qualities, neuroscience, physiology, biology, and medication. Mechanical progressions, for example, hereditary altering apparatuses like CRISPR-Cas9, have empowered analysts to control clock qualities and study their capabilities more meticulously. Devices like optogenetics, which permits the control of cell action with light, have given new roads to examine the complexities of circadian guideline.

Understanding these rhythms not just uncovers the secrets of life on The planet yet in addition holds the way to streamlining human wellbeing, disentangling the intricacies of mental prosperity, and exploring the difficulties of room investigation. As we keep on investigating the wildernesses of science, the investigation of organic rhythms allures us to see the value in the sensitive harmony between the ageless rhythms of the Earth and the steadily developing dance of life.

3.1 Circadian Rhythms: Nature's Daily Clock

Circadian rhythms, frequently portrayed as nature's day to day clock, are an intrinsic element of life on The planet, organizing a wide cluster of physiological and social cycles with an about 24-hour periodicity. The expression "circadian" is gotten from the Latin words "around" (signifying 'around') and "diem" (signifying 'day'), highlighting the cyclic idea of these rhythms.

These inside timekeeping components are found in life forms traversing the whole natural range, from single-celled microorganisms to complex multicellular living beings like people. The omnipresence of circadian rhythms mirrors their essential significance in streamlining a life form's working in synchrony with the World's pivot.

At the core of circadian rhythms is the circadian clock, an inner framework that directs the planning of natural exercises. This endogenous clock permits life forms to expect and adjust to the anticipated changes in their current circumstance related with the World's revolution. While circadian rhythms continue even without a trace of outer signals, they are significantly impacted and synchronized by ecological variables, with the most powerful prompt being the light-dull cycle.

In well evolved creatures, including people, the expert circadian clock lives in the suprachiasmatic core (SCN) of the nerve center. This little area of the cerebrum fills in as the focal pacemaker, organizing and synchronizing fringe timekeepers tracked down in different tissues and organs all through the body. The SCN gets input from the eyes, especially delicate to light, permitting it to adjust the inside circadian musicality to the outside light-dull cycle. This course of synchronization, known as entrainment, guarantees that a living being's inner clock stays as one with the 24-hour day.

The sub-atomic components that underlie circadian rhythms have been a subject of broad examination. At the center of these components are a bunch of "clock qualities" that produce proteins shaping interconnected input circles. In the positive arm of the circle, clock qualities enact the record of different qualities, including those that code for clock proteins. These proteins, thus, repress the action of the clock qualities, making a negative criticism circle. The cyclic ascent and fall of these sub-atomic parts produce the cadenced articulation of qualities related with different physiological cycles, arranging the rhythmic movement of a living being's everyday exercises.

The significance of circadian rhythms is obvious in their unavoidable effect on a large number of organic cycles. One of the most notable circadian rhythms is the rest wake cycle. The circadian clock controls the planning of rest beginning and offset, guaranteeing that times of alertness and rest line up with the outer light-dim cycle. Interruptions to the circadian framework, for example, those accomplished during shift work or transmeridian travel (stream slack), can bring about rest aggravations and effect by and large prosperity.

Past rest, circadian rhythms assume a significant part in the guideline of chemical emission. Chemicals like cortisol, melatonin, and development chemical show circadian examples of delivery, affecting cycles like digestion, insusceptible capability, and the rest wake cycle. The planned arrival of chemicals guarantees that physiological exercises are fittingly coordinated to advance energy use, tissue fix, and other fundamental capabilities.

The circadian framework additionally stretches out its impact to metabolic cycles, influencing the planning of feasts, supplement ingestion, and energy use. Studies have shown that disturbances to circadian rhythms, for example, sporadic eating designs or nighttime eating, can add to metabolic problems and heftiness. The complex interchange between the circadian clock and metabolic pathways highlights the significance of keeping a synchronized day to day everyday practice for in general wellbeing.

Notwithstanding rest, chemical discharge, and digestion, circadian rhythms impact mental capability and temperament. The planning of mental cycles, memory combination, and sharpness is regulated by the circadian framework. Disturbances to circadian rhythms, as found in conditions like shift work, have been related with mental hindrances and mind-set problems. The cozy connection between circadian rhythms and psychological well-being features the many-sided associations between the inside clock and cerebrum capability.

The effect of circadian rhythms stretches out to the cardiovascular framework, impacting circulatory strain, pulse, and the gamble of cardiovascular occasions. Studies have shown that the rate of respiratory failures and strokes follows a circadian example, with a higher probability of these occasions happening in the early morning. The circadian guideline of cardiovascular boundaries underscores the significance of thinking about circadian rhythms in the administration of cardiovascular wellbeing.

In the domain of resistance, circadian rhythms administer the planning of safe reactions, affecting the body's capacity to guard against microbes. The action of invulnerable cells, the development of antibodies, and the fiery reaction all display circadian varieties. Interruptions to circadian rhythms can think twice about capability, possibly expanding vulnerability to contaminations and influencing the viability of immunizations.

The unpredictable dance of circadian rhythms additionally reaches out to the regenerative framework, affecting the planning of richness, periods, and conceptive chemicals. The period in ladies and the hormonal changes related with it follow a circadian example. The circadian guideline of conceptive cycles guarantees that they are synchronized with other physiological exercises, improving the possibilities of effective proliferation.

While the expert circadian clock in the SCN assumes a focal part in organizing circadian rhythms, fringe clocks in different tissues add to the calibrating of musical cycles. These fringe clocks, present in organs like the liver, kidneys, and lungs, are impacted by both the focal clock and nearby ecological signs. The coordination between the focal and fringe timekeepers is fundamental for keeping up with the general cognizance of circadian rhythms all through the body.

The natural prompt that most significantly impacts circadian rhythms is light. Light fills in as a powerful entraining signal, assisting organic entities with changing their interior timekeepers to the outer day-night cycle. The eyes contain photoreceptor cells, especially delicate to light in the blue range, that communicate signs to the SCN, illuminating it regarding the overarching light circumstances. This interaction permits the circadian framework to stay synchronized with the changing seasons and the World's hub slant, guaranteeing that physiological cycles are lined up with the shifting length of sunlight.

The significance of light in directing circadian rhythms has prompted the improvement of light treatment as a treatment for circadian-related messes. Light openness, particularly in the first part of the day, has been demonstrated to be compelling in alleviating the side effects of conditions like occasional emotional problem (Miserable) and certain rest issues. Alternately, limiting openness to splendid light, particularly at night, is prescribed to advance the regular slowing down of the circadian framework in anticipation of rest.

While light is the essential entraining signal, other natural prompts, known as zeitgebers (time-providers), likewise assume a part in directing circadian rhythms. These incorporate factors like temperature, taking care of timetables, and social connections. The consolidated impact of numerous zeitgebers adds to the vigor and versatility of circadian rhythms in various conditions.

The transformative meaning of circadian rhythms is obvious in their preservation across assorted species. From basic creatures like microscopic organisms to complex warm blooded animals, the presence of circadian timekeepers features their versatile worth in advancing endurance and generation. The particular tension inclining toward the development of circadian rhythms is probable connected to the benefits presented by the synchronization of physiological cycles with the anticipated changes in the climate.

The investigation of circadian rhythms has useful ramifications for different fields, including medication, brain research, and chronobiology. Chronobiology, the logical investigation of organic rhythms, looks to grasp the fundamental instruments and utilitarian meaning of circadian and other musical cycles. This interdisciplinary field envelops research in hereditary qualities, neuroscience, physiology, and nature, offering bits of knowledge into the transient association of life.

With regards to clinical applications, the standards of chronobiology have brought about the field of chronotherapy. Chronotherapy includes the streamlining of clinical medicines in light of the circadian timing of physiological cycles. The viability and security of specific meds, as well as the body's reaction to medicines, can shift contingent upon the hour of day. Fitting clinical intercessions to a person's circadian mood holds the possibility to improve treatment results and limit incidental effects.

The investigation of circadian rhythms has additionally extended to investigate their part in different medical issue and sicknesses. Circadian disturbances have been embroiled in the pathogenesis of conditions like diabetes, cardiovascular illness, and malignant growth. Understanding the connections between circadian rhythms and infection is opening roads for the advancement of designated intercessions and customized medication draws near.

As mechanical progressions keep on speeding up, the field of circadian exploration benefits from apparatuses like hereditary altering procedures and high level imaging advances. These devices empower specialists to control clock qualities, research cell and sub-atomic cycles, and envision circadian rhythms at the cell and tissue levels.

3.2 Evolutionary Perspectives on Biological Clocks

The presence of organic clocks, natural timekeeping systems that control the planning of different physiological and social cycles, is well established in the developmental history of life on The planet. From the least difficult single-celled organic entities to complex multicellular living beings, the presence of natural tickers mirrors their versatile importance in enhancing endurance and propagation in a dynamic and cyclic climate. Developmental viewpoints on natural clocks shed light on the different structures and elements of these timekeeping components across the tree of life.

The advancement of natural tickers is personally attached to the World's pivot and its repetitive day-night cycle. The capacity to expect and adjust to cyclic natural changes presented a huge benefit to early living things. The rise of natural timekeepers gave life forms a way to synchronize their inner exercises with the standard variances in light and temperature related with the World's revolution.

The most notable and widely concentrated on natural tickers are circadian rhythms, which have an about 24-hour periodicity. The universality of circadian rhythms across assorted taxa proposes their old beginnings. Indeed, even single-celled organic entities, for example, cyanobacteria, display circadian rhythms in processes like photosynthesis.

The transformative preservation of circadian rhythms highlights their key job in planning organic exercises with the World's turn.

The versatile worth of circadian rhythms is apparent in their job in improving the planning of key organic cycles. For instance, the synchronization of the rest wake cycle with the light-dull cycle is a sign of circadian rhythms in creatures. The capacity to rest during times of obscurity and be dynamic during times of light is profitable for keeping away from hunters, moderating energy, and productively rummaging for food.

With regards to development, the benefits presented by circadian rhythms reach out past individual endurance to affect regenerative achievement. Numerous species show exact timing in their regenerative exercises, adjusting them to positive natural circumstances.

This peculiarity, known as photoperiodism, is controlled by circadian timekeepers and guarantees that regenerative endeavors are maximally fruitful by lining up with ideal natural signs.

The sub-atomic premise of circadian rhythms, explained through hereditary examinations, uncovers a noteworthy level of transformative preservation. Clock qualities, which assume a focal part in controlling circadian rhythms, have homologs across a large number of creatures. The presence of comparative clock qualities in life forms as different as organic product flies, mice, and people recommends that the essential sub-atomic apparatus administering circadian rhythms has antiquated beginnings.

One of the vital bits of knowledge into the development of circadian tickers comes from the investigation of model life forms. In organic product flies (Drosophila melanogaster), the ID of the period quality, a basic part of the circadian clock, established the groundwork for grasping the sub-atomic premise of circadian rhythms. Changes in the period quality outcome in adjusted circadian way of behaving, stressing its fundamental job in the timekeeping system.

Essentially, in warm blooded creatures, the disclosure of the suprachiasmatic core (SCN) as the expert circadian pacemaker featured the significance of brain control in organizing circadian rhythms. The SCN gets input from photoreceptor cells in the eyes, permitting it to synchronize the inside circadian clock with the outer light-dull cycle. This brain control system is monitored across well evolved creatures, highlighting its developmental importance in directing circadian rhythms.

The developmental underlying foundations of circadian clocks stretch out to plants, where they assume a critical part in organizing different physiological cycles. In plants, circadian rhythms impact peculiarities like photosynthesis, supplement take-up, and blossoming. The planning of blossoming, specifically, is a basic variation that guarantees plants recreate when ecological circumstances, like day length and temperature, are helpful for seed improvement.

Past the 24-hour circadian rhythms, longer-term organic rhythms likewise reflect developmental transformations. Occasional rhythms, driven by the World's hub slant and circle around the sun, impact ways of behaving like relocation, hibernation, and propagation. These occasional variations advance a creature's possibilities of endurance and regenerative progress even with changing natural circumstances.

Relocation, saw in many bird species, is a striking illustration of a versatile occasional beat. The planning of relocation lines up with the accessibility of assets and ideal settling conditions. Birds leave on lengthy excursions, navigating mainlands, directed by their inner organic timekeepers receptive to occasional signs. The capacity to explore huge distances with accuracy mirrors the unpredictable exchange between developmental variations and natural tickers.

Hibernation, a condition of lethargy described by decreased metabolic action and energy preservation, is one more occasional variation saw in warm blooded creatures. Sleeping species enter this state during times of natural difficulty, like winter when food is scant. The commencement and end of hibernation are directed by inner natural tickers that answer signs like temperature and day length.

The lunar or flowing rhythms address one more layer of natural timing with transformative importance. The gravitational impact of the moon, combined with its evolving stages, has been proposed to influence the way of behaving of marine life forms. Numerous species show synchronized ways of behaving, like multiplication and movement, connected to the lunar cycle. The versatile benefit of lunar rhythms in marine biological systems mirrors the perplexing exchange between organic tickers and ecological prompts.

The development of organic tickers isn't without its difficulties. Creatures should battle with the continuous need to adjust their inner timekeeping systems to changing ecological circumstances. Developmental tensions, like predation, rivalry for assets, and natural shifts, consistently shape the structure and capability of organic tickers.

The atomic development of clock qualities is dependent upon particular tensions that keep up with their fundamental capabilities while considering transformations to explicit natural specialties. For instance, various species might show varieties in the span of their circadian rhythms, reflecting transformations to their particular natural surroundings and ways of life. Nighttime species might have more limited circadian periods than diurnal species, adjusting their movement designs with the overarching states of obscurity or light.

The development of natural tickers is additionally unpredictably connected to biological communications and interspecies connections. Hunters and prey, for example, may show corresponding circadian rhythms that enhance their possibilities experiencing or staying away from one another. The synchronization of natural tickers inside environments adds to the general strength and working of biological networks.

The investigation of transformative points of view on organic clocks has down to earth suggestions for understanding the effects of natural changes, including environmental change and living space interruption. Changes in ecological signs, for example, modifications in temperature, precipitation examples, and day length, can upset the synchronization of organic timekeepers with their normal rhythms. These disturbances might prompt jumbles in regenerative timing, movement designs, and other basic life processes, with expected ramifications for populace elements and species endurance.

With regards to human advancement, the appearance of fake lighting and current ways of life has acquainted new difficulties with our circadian rhythms. The broad utilization of electric lighting, shift work, and steady openness to screens with fake light has established conditions

that can disturb the regular synchronization of our circadian timekeepers. The subsequent circadian misalignment has been related with different medical problems, including rest issues, metabolic aggravations, and an expanded gamble of constant infections.

As we dive into the complexities of transformative points of view on organic clocks, it becomes clear that these timekeeping components are not static substances but rather unique variations that have developed north of millions of years. The transaction between hereditary variables, natural signs, and specific tensions has molded the different structures and elements of organic tickers across the tree of life. From the earliest life structures to the mind boggling environments we notice today, the developmental story of organic clocks is a demonstration of the wonderful manners by which life has adjusted to the musical rhythm of the Earth.

3.3 Geological Time: Rocks, Fossils, and Earth's Epochs

Geographical time, the tremendous spread of time crossing the World's set of experiences, is a material whereupon the unpredictable accounts of rocks, fossils, and ages are painted. The topographical record fills in as a narrative of Earth's development, reporting the powerful cycles that have formed the planet more than billions of years. From the arrangement of rocks to the development of life and the moving ages that mark huge land occasions, the investigation of geographical time reveals the profound history of our planet.

Rocks, the structure blocks of the World's outside layer, convey inside them the engravings of topographical cycles and occasions. The investigation of rocks, known as petrology, gives urgent experiences into the World's set of experiences, including arrangement and the different powers have formed its surface. Rocks fall into three principal classifications in light of their arrangement processes: volcanic, sedimentary, and transformative.

Volcanic rocks structure from the cooling and cementing of liquid magma. The pace of cooling and the mineral piece decide the surface of the stone. Stone, with its coarse-grained surface, is an illustration of a volcanic stone that structures gradually underneath the World's surface. Conversely, basalt, with its fine-grained surface, results from quick cooling at the surface.

Sedimentary rocks, then again, are framed through the gathering and cementation of residue. These dregs can incorporate particles of minerals, natural material, and, surprisingly, the remaining parts of life forms. After some time, layers of residue go through compaction and cementation to become rocks like sandstone, limestone, and shale. Fossils, essential markers of Earth's set of experiences, are in many cases tracked down safeguarded in sedimentary rocks.

Transformative rocks go through changes because of changes in temperature, pressure, or the presence of artificially dynamic liquids. Previous rocks, whether volcanic, sedimentary, or other transformative rocks, can be exposed to these changes, bringing about the arrangement of

rocks like marble, schist, and record. The course of transformation changes the mineral structure and surface of the stones, abandoning proof of the topographical powers at play.

Fossils, implanted in sedimentary rocks like a depiction of old life, give a window into the World's past. These safeguarded remains or hints of creatures offer important experiences into the advancement of life throughout land time. Fossilization happens when the remaining parts of plants and creatures are covered and shielded from rot by dregs. The progressive supplanting of natural material with minerals brings about the arrangement of fossils.

Fossils come in different structures, from bones and shells to engravings and tracks. The investigation of fossils, known as fossil science, permits researchers to reproduce the World's organic history, uncovering the variety of previous existence structures and the developmental cycles that have formed the tree of life. Fossils give proof of wiped out species as well as deal pieces of information about old biological systems, environment conditions, and the land occasions that impacted life on The planet.

The idea of profound time, first expressed by Scottish geologist James Hutton in the eighteenth hundred years, changed the comprehension of Earth's set of experiences. Profound time alludes to the limitlessness of geographical time, reaching out more than billions of years. Hutton's standard of uniformitarianism, which expresses that similar normal cycles noticed today have worked over Earth's time, laid the basis for understanding the enormous timescales engaged with land processes.

Geographical time is coordinated into a progressive structure that incorporates ages, times, periods, ages, and ages. The biggest unit, the age, addresses the longest divisions of land time. Earth's set of experiences is ordinarily partitioned into four ages: the Hadean, Archean, Proterozoic, and Phanerozoic. The Phanerozoic age, the latest and progressing, incorporates the advancement of complicated living things and is additionally partitioned into times.

The Paleozoic, Mesozoic, and Cenozoic times are significant divisions of the Phanerozoic age. Every period is described by particular topographical and natural occasions. The Paleozoic period, known as the time of old life, saw the rise of different marine spineless creatures, the colonization of land by plants and creatures, and the advancement of early vertebrates. The conclusion of the Paleozoic age was set apart by the biggest mass annihilation occasion in Earth's set of experiences, prompting the vanishing of various species.

The Mesozoic period, frequently alluded to as the time of dinosaurs, ranges from around 252 to quite a while back. Dinosaurs, which ruled earthbound biological systems during this period, developed into a wide cluster of structures and sizes.

The Mesozoic time additionally saw the ascent of warm blooded creatures, the advancement of birds, and massive changes in vegetation. The conclusion of the Mesozoic age is set apart by one more mass annihilation occasion, which prompted the termination of dinosaurs and made ready for the expansion of vertebrates.

The Cenozoic period, the time of late life, started around quite a while back and proceeds to the current day. It is described by the further advancement of warm blooded animals, the extension of fields, the improvement of current environments, and the rise of people. The Cenozoic period is separated into ages, including the Paleogene, Neogene, and Quaternary ages, each undeniable by particular topographical and organic occasions.

The Quaternary age, the latest age of the Cenozoic period, ranges the last 2.6 million years and is described by the beginning of icy interglacial cycles. The Quaternary age is set apart by the presence of Homo sapiens, the improvement of complicated human social orders, and the huge effect of people on the World's current circumstance. The continuous Holocene age, which started something like a long time back, is a development of the Quaternary age.

Land time is additionally partitioned into ages, which address more limited timespans portrayed by unambiguous topographical and natural occasions. Ages are characterized in view of the event of particular stone layers, fossils, and other geographical markers. The Holocene age, for instance, is partitioned into three ages: the Greenlandian, the Northgrippian, and the Meghalayan, each undeniable by unambiguous climatic and natural changes.

The idea of geographical time is principal to figuring out Earth's set of experiences, and different dating methods permit researchers to appoint mathematical ages to rocks and fossils. Radiometric dating, which depends on the rot of radioactive isotopes, is an essential strategy for deciding the outright periods of rocks and minerals. Scientifically measuring, a particular type of radiometric dating, is utilized to decide the time of natural materials up to roughly 50,000 years of age.

Geographical time fills in as a setting for understanding Earth's dynamic cycles, including plate tectonics, environmental change, and the development of life. Plate tectonics, the development of Earth's lithospheric plates, brings about the arrangement of mountain reaches, seismic tremors, and the formation of new maritime hull. The moving landmasses and the opening and shutting of sea bowls are apparent in the geographical record, giving experiences into the World's steadily evolving surface.

Environmental change, a repetitive subject over Earth's time, is reported in the geographical record through proof like icy stores, fossilized dust, and changes in sedimentation designs. The investigation of past environment varieties gives setting to understanding current environmental change and its likely effects on biological systems and human social orders.

Mass termination occasions, set apart by the quick decay of a huge part of Earth's biodiversity, are key occasions in land time. The fossil record jelly proof of a few mass eliminations, including the end-Ordovician, end-Devonian, end-Permian, end-Triassic, and end-Cretaceous terminations. Every one of these occasions significantly affected the advancement of life and the ensuing enhancement of enduring species.

The end-Cretaceous mass elimination, which happened around a long time back, is one of the most notable eradication occasions. It prompted the end of the dinosaurs and numerous different species, possible set off by the effect of a huge space rock or comet. The fallout of this occasion made ready for the ascent of warm blooded animals and the inevitable development of people.

The investigation of geographical time keeps on being a powerful field of logical request. Progresses in innovation, including modern dating methods, remote detecting, and PC displaying, upgrade our capacity to translate Earth's set of experiences with expanding accuracy. The incorporation of information from different logical disciplines, like geography, fossil science, climatology, and geochemistry, permits scientists to build itemized accounts of Earth's advancement.

As we peer into the profundities of geographical time, we gain a significant comprehension of Earth's set of experiences as well as bits of knowledge into the cycles that have formed the planet and the interconnectedness of land, natural, and climatic peculiarities. The continuous mission to unwind the secrets of land time is an investigation of the World's past, present, and future — an excursion that reaches out back billions of years and keeps on unfurling with each layer of rock, each fossil, and each age that is uncovered.

3.4 Impact of Human Activity on Earth's Temporal Landscape

The effect of human movement on Earth's worldly scene is an unfurling account that traverses hundreds of years yet has advanced rapidly in the cutting edge period. As stewards of the planet, people have made a permanent imprint on the worldly rhythms of Earth, influencing everything from environment examples and biological systems to land processes. This effect, driven by innovative headways, industrialization, and populace development, has significant ramifications for the planet's fleeting elements and the sensitive equilibrium of its regular frameworks.

Quite possibly of the main manner by which human action adjusts Earth's fleeting scene is through environmental change. The consuming of non-renewable energy sources, deforestation, and modern cycles discharge enormous amounts of ozone depleting substances, like carbon dioxide and methane, into the air. These gases trap heat, prompting a warming of the planet — an impact generally alluded to as an Earth-wide temperature boost.

The outcomes of environmental change are expansive and influence different transient parts of Earth's frameworks. One of the most perceptible effects is the change of weather conditions and the strengthening of outrageous occasions. Changes in temperature, precipitation, and tempest recurrence upset laid out climatic standards, affecting the transient rhythms of seasons, developing seasons, and normal cycles.

Climbing worldwide temperatures add to the softening of polar ice covers and glacial masses, influencing ocean levels and sea flows. The worldly elements of sea dissemination, pivotal for controlling environment, are upset by the deluge of freshwater from softening ice. Changes in ocean surface temperatures impact the recurrence and power of occasions like tropical storms and hurricanes, adjusting the worldly examples of these super climate peculiarities.

The transient outcomes of environmental change stretch out to biological systems and biodiversity. Changes in temperature and precipitation designs upset the occasional ways of behaving of plants and creatures. Phenological occasions, like the planning of blooming, movement, and multiplication, are firmly connected to environment conditions. As the environment changes, the fleeting synchrony between communicating species can be disturbed, affecting natural connections and the working of biological systems.

Human-incited environmental change likewise presents dangers to transient scenes through sea fermentation. The retention of overabundance carbon dioxide by the seas brings about expanded corrosiveness, influencing marine existence with calcareous skeletons or shells. Coral reefs, fundamental biological systems with mind boggling transient elements, are especially powerless. The corruption of coral reefs upsets the worldly examples of biodiversity, influencing the incalculable species that rely upon these environments.

Land-use changes driven by human exercises further shape Earth's transient scene. Deforestation, urbanization, and rural extension adjust the fleeting elements of biological systems, prompting environment misfortune, fracture, and debasement. The change of regular scenes into human-ruled conditions disturbs the transient rhythms of species adjusted to explicit natural specialties, frequently prompting decreases in biodiversity.

The transient results of land-use changes reach out to soil disintegration, supplement cycling, and water accessibility. Deforested regions are more vulnerable to disintegration, changing the fleeting examples of silt transport and affidavit in waterways and seas. Changes in land cover impact the accessibility of water assets, influencing the fleeting elements of stream and groundwater re-energize. The change of regular scenes has flowing impacts on the worldly associations inside biological systems.

The extraction and utilization of regular assets, driven by human interest, add to the change of Earth's worldly scene. The double-dealing of petroleum derivatives, minerals, and freshwater assets happens at rates that surpass the planet's ability for recovery. The extraction of groundwater, for instance, can prompt the consumption of springs, modifying the worldly elements of water accessibility and adding to long haul ecological debasement.

Mining exercises, one more illustration of asset extraction, can lastingly affect topographical and worldly scenes. Open-pit mining, deforestation, and the arrival of toxins into soil and water adjust the transient examples of geographical cycles. The scars left by mining exercises upset normal transient rhythms, influencing soil ripeness, water quality, and the general strength of biological systems.

The utilization of innovation in farming has likewise changed Earth's transient scene. The far and wide reception of modern cultivating works on, including the utilization of composts, pesticides, and monoculture, has expanded horticultural efficiency yet has come at the expense of ecological debasement. The interruption of normal worldly rhythms, like yield turn and the decrepit periods fundamental for soil recovery, can prompt soil corruption, loss of biodiversity, and supplement irregular characteristics.

The transient outcomes of concentrated farming reach out to water assets. The utilization of water system, fundamental for supporting yields in bone-dry areas, can prompt the exhaustion of water sources and the modification of stream designs. Changes in the transient elements of water accessibility influence sea-going environments, amphibian biodiversity, and the occupations of networks reliant upon water assets.

The fleeting scene of Earth is likewise impacted by the far reaching utilization of manufactured synthetics in different enterprises. The arrival of poisons out of sight, water, and soil lastingly affects natural wellbeing. Diligent natural poisons, weighty metals, and other harmful substances can gather in biological systems, upsetting the transient examples of supplement cycling, sullying pecking orders, and presenting dangers to human and environmental wellbeing.

The remarkable development of human populaces escalates the effect on Earth's fleeting scene. The requests for food, water, energy, and living space have prompted phenomenal paces of asset utilization and waste age. The transient outcomes of populace development are obvious in the consumption of regular assets, the deficiency of biodiversity, and the gathering of toxins in the climate.

The idea of the Anthropocene, a proposed age characterized by the huge effect of human exercises on Earth's geography and environments, mirrors the affirmation of people as a significant land force.

The transient scene of the Anthropocene is portrayed by sped up changes in environment, biodiversity misfortune, modified supplement cycles, and disturbances to normal fleeting rhythms. The acknowledgment of the Anthropocene highlights the requirement for reasonable practices and a reconsideration of the human relationship with the planet.

Endeavors to relieve the effect of human movement on Earth's transient scene include a multidisciplinary approach enveloping science, strategy, and cultural activities. Feasible practices that think about the transient elements of biological systems, environment, and asset use are fundamental for protecting the planet's wellbeing and versatility. Preservation drives, reforestation projects, and the security of normal territories add to the rebuilding of fleeting rhythms inside biological systems.

The turn of events and reception of environmentally friendly power advancements assume an essential part in relieving the effect of human exercises on Earth's worldly scene. Progressing away from petroleum products to inexhaustible sources, for example, sunlight based, wind, and hydropower, diminishes ozone harming substance outflows and reduce the effect of environmental change. Embracing economical energy rehearses adds to the rebuilding of worldly equilibrium in Earth's environment frameworks.

The execution of maintainable horticulture rehearses is imperative for reestablishing the transient scene of biological systems. Agroecological approaches, including natural cultivating, agroforestry, and crop enhancement, advance the recovery of soil wellbeing, the protection of biodiversity, and the strength of agrarian frameworks to environmental change. Feasible farming practices add to the reclamation of regular worldly rhythms inside scenes.

Ecological strategies and guidelines that focus on preservation, contamination counteraction, and maintainable asset the board are fundamental for relieving the effect of human exercises on Earth's worldly scene. Peaceful accords, for example, the Paris Settlement on environmental change and the Show on Natural Variety, highlight the worldwide acknowledgment of the requirement for aggregate activity to address ecological difficulties and safeguard the planet's transient elements.

Instructive drives and public mindfulness crusades assume a basic part in encouraging a more profound comprehension of the interconnectedness between human exercises and Earth's worldly scene. Empowering maintainable ways of behaving, capable utilization, and natural stewardship enables people and networks to add to the rebuilding and conservation of the planet's worldly rhythms.

The effect of human movement on Earth's fleeting scene is a perplexing and complex test that requires worldwide participation, inventive arrangements, and a guarantee to supportable practices.

As mankind explores Anthropocene, the decisions today will shape the worldly elements of Earth for a long time into the future. Perceiving the interconnectedness of normal frameworks and embracing an agreeable relationship with the planet are fundamental stages toward reestablishing and supporting the transient scene of our common home.

Chapter 4

Quantum Time and the Microscopic Universe

Quantum mechanics has for some time been a wellspring of interest and bewilderment, testing how we might interpret the principal idea of the real world. One of the most charming parts of this field is the idea of quantum time, which investigates the idea of time at the infinitesimal level. In the domain of the tiny, where particles dance and cooperate in manners that resist traditional instinct, time assumes a particular personality that requests a reexamination of our regular thoughts.

At the core of quantum mechanics is the wave-molecule duality, a peculiarity that features the double idea of particles like electrons and photons. These substances show both wave-like and molecule like properties, contingent upon the trial conditions. This duality is embodied by the popular twofold cut try, where particles terminated through two cuts make an impedance design normal for waves. However, when the particles are noticed, they act like discrete particles, apparently mindful of being watched.

The wave-molecule duality challenges our instinctive comprehension of the real world, provoking physicists to dive further into the idea of the quantum world. In doing as such, they have experienced the mystery of quantum time, an idea that veers off from our regular experience of time as a persistent and straight movement.

In traditional physical science, time is a flat out and unidirectional element, streaming consistently from the past to the present and into what's in store. Notwithstanding, the quantum domain presents a degree of vulnerability and non-linearity that withdraws from this old style structure. The idea of time in the minute universe is definitely not a direct bolt however a perplexing exchange of probabilities and possibilities.

One critical element of quantum time is the job of perception and estimation. In the quantum domain, the demonstration of estimation is definitely not a uninvolved perception however a functioning commitment that impacts the framework being noticed. This is exemplified by Heisenberg's vulnerability rule, which expresses that the more definitively we know a molecule's situation, the less exactly we can know its force, as well as the other way around. The demonstration of estimation upsets the framework, bringing an innate vulnerability into the actual texture of the quantum world.

The idea of time, as well, becomes caught with the demonstration of estimation. The demonstration of noticing a quantum framework at a particular second appears to implode the scope of possibilities into a solitary fact, prompting whether or not time, in the quantum domain, is a ceaseless stream or a progression of discrete minutes directed by perception.

One understanding of quantum time is the "block universe" model, where past, present, and future all coincide in a static block. In this view, the whole history of the universe is spread out, and the progression of time is a deception. Each second in time is similarly genuine, and our impression of time elapsing is a result of our emotional experience.

Nonetheless, the block universe model brings up philosophical issues about through and through freedom and determinism. In the event that what's in store is now foreordained, do we really have the opportunity to simply decide, or would we say we are basically entertainers following a content composed by the changeless laws of material science? The ramifications of this model stretch past the limits of material science, digging into the domains of reasoning and power.

One more captivating part of quantum time is the peculiarity of time widening, a result of Einstein's hypothesis of relativity. As particles approach the speed of light, time for them seems to dial back comparative with eyewitnesses very still. This relativistic impact has been tentatively affirmed and is a fundamental consider how we might interpret the texture of spacetime.

The interaction between quantum mechanics and relativity presents a rich embroidery of thoughts and difficulties. The journey for a brought together hypothesis that can consistently blend the minute universe of quantum mechanics with the inestimable scales depicted by broad relativity stays one of the sacred goals of contemporary material science.

String hypothesis, a hypothetical structure that places the presence of one-layered "strings" as the major structure blocks of the universe, has arisen as a possibility for such a brought together hypothesis. In string hypothesis, the texture of spacetime isn't nonstop yet made out of small, vibrating strings. These strings lead to the particles and powers saw in the universe, offering an enticing look into a more major layer of the real world.

The tiny universe, as uncovered by quantum mechanics, challenges how we might interpret time as well as our ideas of causality. In the quantum domain, circumstances and logical results become entwined in manners that challenge traditional thinking. The idea of trap, where particles become related so that the condition of one momentarily impacts the condition of the other, no matter what the distance between them, embodies this takeoff from traditional causality.

Snare has been tentatively checked through trial of Chime's disparities, which demonstrate the way that the connections between's trapped particles can't be made sense of by old style physical science. The ramifications of entrapment reach out past the limits of the lab, starting discussions about the idea of the real world and the interconnectedness of the universe.

The peculiarity of ensnarement brings up the issue of whether there exists a secret layer of the real world, frequently alluded to as "stowed away factors," that decides the results of quantum estimations. While certain translations of quantum mechanics place the presence of such secret factors, others, similar to the Copenhagen understanding, accentuate the job of perception in characterizing reality.

The philosophical discussions encompassing the translation of quantum mechanics feature the significant and bewildering nature of the tiny universe. The inborn vulnerabilities, probabilities, and non-neighborhood associations challenge our instincts and push the limits of what we can get a handle on about the principal idea of the real world.

In the mission to comprehend quantum time and the minute universe, researchers and scholars the same have investigated the idea of spacetime itself. The idea of spacetime, presented by Einstein's hypothesis of relativity, brings together reality into a solitary, four-layered continuum. In this structure, gravity isn't a power communicated through space yet an ebb and flow of spacetime brought about by the presence of mass and energy.

Quantum field hypothesis, one more mainstay of present day physical science, expands this thought by portraying particles as excitations of fundamental quantum handles that saturate spacetime. The marriage of quantum mechanics and relativity in the system of quantum field hypothesis has been colossally fruitful in making sense of the way of behaving of particles and their cooperations.

Nonetheless, the marriage of quantum mechanics and relativity isn't without its difficulties, especially with regards to understanding the idea of spacetime at the littlest scales. The granularity of spacetime itself turns into a subject of examination, with questions emerging about whether spacetime is constant or made out of discrete structure blocks.

A hypotheses propose a principal granularity to spacetime, recommending that at the Planck scale — generally $10^{(-35)}$ meters — spacetime is certainly not a smooth continuum yet a matrix like design. This granularity presents a key unit of length and time, testing our traditional ideas of smooth, nonstop space.

Circle quantum gravity is one such hypothesis that investigates the discrete idea of spacetime. In this system, spacetime is quantized, and the texture of the universe is woven from interconnected circles. The quantization of spacetime at the littlest scales opens up new roads for grasping the idea of the quantum domain and its suggestions for the central design of the real world.

As physicists dig into the minuscule universe and wrestle with the complexities of quantum time, they are defied with the limits of our ongoing applied systems. The very language we use to depict the quantum world might be deficient, and new numerical formalisms and applied apparatuses might be expected to disentangle its secrets.

One such road of investigation is the idea of quantum data. Quantum data hypothesis, a part of quantum mechanics that regards data as an actual element, has given new bits of knowledge into the idea of quantum frameworks. The possibility that data is central to the texture of reality challenges our customary perspectives on issue and energy as the structure blocks of the universe.

Quantum trap, a peculiarity personally attached to the idea of quantum data, has been investigated with regards to quantum figuring. The potential for utilizing entrapment to perform calculations at speeds unreachable by old style PCs opens up new vistas in the domain of data handling.

The investigation of quantum data has likewise prompted the proposition of the holographic rule, an idea recommending that the data content of a three-layered volume can be encoded on a two-layered surface encompassing it. This rule, propelled by dark opening physical science, indicates a more profound association between quantum mechanics and gravity, testing how we might interpret spacetime itself.

The holographic rule brings up significant issues about the idea of the real world and the connection between the infinitesimal and naturally visible scales. On the off chance that the data content of the universe is encoded on a two-layered surface, what does this infer for our view of three-layered space? Does the texture of reality have a more profound, holographic construction that evades our immediate discernment?

The secrets of quantum time and the minuscule universe likewise meet with the vast story of the universe's beginnings and development. The area of cosmology, which looks to grasp the huge scope design and elements of the universe, is personally associated with the experiences gathered from quantum mechanics and relativity.

The inestimable microwave foundation radiation, a weak gleam left over from the early snapshots of the universe, gives a depiction of the universe when it was only two or three hundred thousand years of age. The examples and variances in this radiation encode critical data about the seeds of astronomical construction and the advancement of the universe north of billions of years.

The transaction between quantum variances at the minuscule level and the enormous development of the universe is a captivating embroidery of interconnected peculiarities. Quantum variances in the early universe are remembered to have brought about the designs we notice today — systems, groups of worlds, and tremendous enormous web-like designs.

As the universe extended and cooled, quantum variances were extended to vast scopes, leaving engraves for the enormous scope design of the universe. The investigation of these engravings, whether through perceptions of the grandiose microwave foundation or the dispersion of cosmic systems, gives a novel window into the quantum beginnings of the universe.

The enormous inflationary hypothesis, which places a quick dramatic development of the universe in its initial minutes, is one more feature of the grandiose story impacted by the minuscule domain. Quantum changes during the inflationary age are remembered to have cultivated the thickness irritations that ultimately developed into the designs we see in the universe today.

While the inestimable story gives an excellent account of the universe's development for huge scopes, the tiny universe presents a degree of intricacy and vulnerability that resounds through infinite history. The journey to comprehend the associations between the quantum and astronomical domains keeps on moving analysts to push the limits of our insight.

Chasing a brought together hypothesis that can flawlessly include the tiny and vast scopes, scientists wrestle with the considerable difficulties presented by the intrinsic intricacies of both quantum mechanics and cosmology. The slippery idea of dull matter and dim energy, which together comprise around 95% of the universe, adds one more layer of secret to the vast story.

Dim matter, a type of issue that doesn't interface with light and stays undetectable to telescopes, applies a gravitational impact on noticeable matter. The idea of dim matter particles is one of the remarkable riddles in contemporary material science, and their disclosure would have significant ramifications for how we might interpret the central constituents of the universe.

Dull energy, then again, is a strange power driving the sped up extension of the universe. The idea of dull energy stays quite possibly of the main unsettled question in cosmology, with suggestions for a definitive destiny of the universe.

The journey for a bound together hypothesis that can represent the minuscule and enormous secrets has prompted the investigation of intense thoughts and speculative structures. The idea of a multiverse, where our universe is only one of numerous universes with various actual constants and properties, has gotten forward momentum in hypothetical material science.

The multiverse speculation emerges from the acknowledgment that the crucial constants of nature, like the speed of light or the strength of gravity, are finely tuned to consider the presence of mind boggling structures like systems and stars. In a multiverse situation, the thought is that various districts of room could have various qualities for these constants, and our universe is in a locale helpful for the development of life.

While the multiverse theory is speculative and as of now needs direct observational proof, it highlights the significant secrets and interconnectedness of the tiny and inestimable domains. The mission for a brought together hypothesis that can exquisitely make sense of the texture of reality at all scales keeps on being a main impetus in hypothetical physical science.

The difficulties presented by the minuscule universe reach out past the limits of conventional physical science and into the domains of reasoning and mysticism. The actual idea of the real world, the presence of stowed away layers past our nearby insight, and the transaction among cognizance and the actual world are subjects that overcome any barrier among science and reasoning.

The job of the onlooker in quantum mechanics has for quite some time been a subject of philosophical thought. The demonstration of estimation, which appears to fall the scope of possibilities into a solitary fact, brings up issues about the idea of perception and its part in characterizing reality. Does the demonstration of perception make reality, or does it only uncover prior parts of the quantum world?

The connection among cognizance and the minute universe is a subject of progressing investigation and discussion. A few scholars recommend that cognizance assumes a key part in forming the idea of the real world, proposing a more profound association between the eyewitness and the noticed. Others contend for a more reductionist view, declaring that cognizance is a new property of complicated brain processes.

The mystery of quantum time adds one more layer to the philosophical contemplating. In the event that time in the quantum domain is certainly not a persistent stream however a progression of discrete minutes directed by perception, what does this suggest for how we might interpret the idea of presence? Does the idea of a block universe, where past, present, and future coincide, challenge our thoughts of choice and organization?

As we explore the intricacies of quantum time and the infinitesimal universe, the mission for understanding reaches out to the actual underpinnings of our perspective. The investigation of these significant inquiries requires a blend of logical request, philosophical consideration, and an eagerness to push the limits of our ongoing comprehension.

4.1 Quantum Entanglement: A Timeless Connection

Quantum snare remains as quite possibly of the most captivating and confounding peculiarity in the domain of quantum mechanics. It challenges our old style instincts about the separateness of items and presents an idea of interconnectedness that rises above the limits of reality. At its center, ensnarement uncovers an immortal association between particles, testing our ordinary comprehension of the key idea of the real world.

The peculiarity of entrapment emerges when at least two particles become corresponded so that the condition of one molecule is straightforwardly connected with the condition of another, no matter what the distance between them. This connection endures in any event, when the particles are isolated by tremendous grandiose distances, recommending a type of prompt correspondence that appears to disregard the vast speed limit forced by the speed of light.

The trial confirmation of trap traces all the way back to the renowned EPR (Einstein-Podolsky-Rosen) mystery proposed in 1935. Albert Einstein, Boris Podolsky, and Nathan Rosen expected to challenge the culmination of quantum mechanics by featuring what they saw as "creepy activity a good ways off." That's what they contended assuming that the properties of one ensnared molecule were estimated, the condition of the other molecule, regardless of the distance away, would still up in the air.

This obvious non-region innate in trap grieved Einstein, who broadly alluded to it as "creepy activity a good ways off." He found it challenging to accommodate with the standards of relativity, which place that no data or impact can travel quicker than the speed of light. Be that as it may, resulting tests, for example, the Chime tests, affirmed the presence of trap and tested old style instincts about the idea of the real world.

One of the original examinations approving the truth of ensnarement is the Viewpoint analyze, led by physicist Alain Perspective during the 1980s. The analysis utilized caught photons and estimated their polarizations this way and that.

The consequences of the analysis abused Chime's imbalances, giving solid proof for trap and precluding specific types of stowed away factor hypotheses that looked to traditionally make sense of quantum relationships.

The snare peculiarity, albeit tentatively checked, remains covered in secret. The system by which caught particles impart or share data across immense distances stays subtle. This puzzler brings up significant issues about the idea of spacetime, the job of perception, and the crucial interconnectedness of the universe.

One understanding of ensnarement includes the idea that the trapped particles exist in a common quantum state, depicted by a bound together wave capability. In this view, the demonstration of estimating one molecule quickly influences the condition of the other, no matter what the spatial partition. The test, be that as it may, lies in understanding how such prompt relationships can happen without disregarding the standards of relativity.

One more translation of ensnarement includes the idea of non-neighborhood stowed away factors. As per this thought, the entrapped particles have specific secret properties or factors that decide their results when estimated. These secret factors, thus, consider the apparently quick connection between's estimations. Be that as it may, this translation faces difficulties and discussions, remembering the infringement of Chime's imbalances for trial tests.

The idea of entrapment has found commonsense applications in arising advances, especially in the field of quantum data and quantum figuring. Quantum ensnarement empowers the production of snared qubits, the essential units of quantum data. Changes in the condition of one qubit promptly influence the condition of its snared accomplice, giving an extraordinary asset to quantum data handling.

Entrapment based quantum key appropriation is one more application with huge ramifications for secure correspondence. By utilizing snared particles, quantum correspondence frameworks can empower secure key dispersion, where any listening in endeavors would upset the trapped states and be recognizable. This component use the one of a kind quantum properties of ensnarement for the improvement of quantum correspondence conventions that are innately secure.

The investigation of quantum entrapment has likewise prompted the proposition of trap helped innovations for undertakings like quantum metrology and detecting. The responsiveness and accuracy accomplished in estimations utilizing caught states outperform the abilities of old style techniques, opening new roads for progressions in logical instrumentation and mechanical applications.

While the functional uses of entrapment feature its true capacity for reforming data handling and correspondence, the central inquiries encompassing its temperament continue. The puzzling momentary relationship between's trapped particles challenges our instinctive comprehension of causality and prompts a reexamination of the idea of reality at the quantum level.

The ageless association suggested by trap brings up interesting issues about the idea of time itself. In the snared domain, the thought of past, present, and future assumes an alternate personality. The relationships between's snared particles appear to rise above the requirements of time as we see it in our regular encounters.

The idea of a block universe, where past, present, and future exist together as static components, tracks down reverberation in the snare peculiarity. In the event that the condition of one snared molecule is momentarily impacted by the estimation of its accomplice, no matter what the spatial division, the customary straight movement of time turns into a more perplexing exchange of connections.

The trap of time and quantum states presents a degree of non-region that challenges our traditional comprehension of worldly request. In the ensnared domain, the future estimation of one molecule appears to impact the previous condition of its snared accomplice, making an immortal association that rises above the bolt of time as we see it.

The philosophical ramifications of ensnarement stretch out past the logical domain, addressing inquiries concerning the idea of the real world, causality, and the constraints of our insight. The ensnared universe challenges our old style instincts and coaxes us to investigate the outskirts of our comprehension.

The interconnectedness uncovered by snare additionally brings up issues about the idea of cognizance and its part in the quantum world. The demonstration of estimation, which implodes the superposition of quantum states and decides the result, presents a component of eyewitness reliance. This spectator impact has provoked hypothesis about the job of cognizance in molding reality at the quantum level.

A few understandings of quantum mechanics recommend that cognizance assumes a key part in the estimation cycle, suggesting a profound association between the spectator and the noticed. This thought, frequently connected with the Copenhagen understanding, places that the demonstration of perception brings the quantum world into a clear state.

Notwithstanding, the connection among cognizance and quantum peculiarities stays a subject of discussion and investigation. While certain scholars propose a more dynamic job for cognizance in the quantum domain, others advocate for a more impartial, eyewitness free comprehension of quantum processes.

The investigation of trap and its suggestions for the idea of reality has prompted the investigation of more extensive grandiose inquiries. The idea of entrapment reverberations through the grandiose snare of huge scope structures, where universes and world bunches are interconnected by the gravitational powers that tight spot them together.

The infinite web, a huge organization of fibers and voids that characterizes the enormous scope construction of the universe, displays an example suggestive of entrapment on inestimable scales. The gravitational collaborations between vast designs make relationships and associations that rise above the limits of individual worlds, forming the inestimable woven artwork on the biggest scales.

The investigation of infinite trap reaches out to the investigation of inestimable microwave foundation radiation (CMB), the weak shine left over from the early snapshots of the universe. The examples and variances in the CMB encode data about the seeds of grandiose design and the quantum vacillations that led to the enormous scope structures noticed today.

The entwined stories of quantum trap and grandiose associations welcome us to consider a universe where the strings of interconnectedness stretch from the quantum domain to the inestimable scales. The investigation of these associations provokes us to rethink our ideas of room, time, and the crucial idea of the real world.

In the journey to comprehend ensnarement and its suggestions, scientists are pushing the limits of both quantum mechanics and cosmology. The quest for a bound together hypothesis that can consistently incorporate the minute and vast scopes keeps on rousing intense thoughts and hypothetical systems.

The difficulties presented by entrapment likewise reach out to the advancement of novel trials and innovations pointed toward tackling its interesting properties. Quantum entrapment, when thought about a confounding and irrational part of quantum mechanics, is currently a significant asset for the progression of quantum innovations with applications going from secure correspondence to quantum figuring.

4.2 Time Crystals: Breaking Symmetry in Quantum Systems

Time gems, a captivating and moderately late idea in the domain of quantum physical science, challenge how we might interpret balance, time, and the principal idea of issue. These exceptional designs address a clever period of issue that breaks both spatial and fleeting balances in a manner recently remembered to be unthinkable. The investigation of time precious stones has opened up new roads for figuring out the quantum world and can possibly reform regions, for example, quantum registering and the underpinnings of thermodynamics.

The idea of time precious stones started in the brain of Nobel laureate Straightforward Wilczek, who proposed the thought in 2012. Conventional gems show a rehashing spatial example, for example, the customary game plan of particles in a jewel or the grid design of a snowflake. Conversely, time precious stones show a rehashing worldly example, wavering between various states without the contribution of outer energy.

At the core of time gems is the idea of unconstrained evenness breaking, a peculiarity that assumes a critical part in the way of behaving of issue at both the quantum and traditional levels. Evenness breaking happens when a framework in a symmetric state changes to a less symmetric or totally deviated state, frequently bringing about the development of new and surprising properties.

On account of time gems, the breaking of time balance is especially interesting. In a standard gem, particles orchestrate themselves in a rehashing design in space, framing a construction that looks the equivalent no matter what the heading in which it is seen. Time gems, nonetheless, display a rehashing design in time, going through changes that happen at normal stretches, breaking the evenness of time itself.

The primary hypothetical proposition for a period precious stone imagined a process for collaborating turns in a spatial grid. The thought was to have these twists waver between two states in a way that rehashes occasionally, even without outer powers. This unconstrained wavering, breaking both spatial and worldly balances, was hypothesized to be a mark of the slippery time gem stage.

In 2016, specialists from Microsoft and Princeton College, drove by Christopher Monroe and Norman Yao, separately, took huge steps in acknowledging time precious stones in the research center. They involved a chain of ytterbium particles as the reason for their investigation, utilizing accuracy strategies to control the quantum conditions of these particles. The subsequent framework displayed the trademark highlights of a period precious stone, with the particles' attractive minutes swaying in an occasional design without the requirement for consistent outer information.

The acknowledgment of time gems in the research facility denoted a noteworthy accomplishment, affirming that these novel periods of issue could for sure exist. The analysis showed the way that time gems could rise up out of the intrinsic elements of quantum frameworks, testing the conventional comprehension of balance and the security of such stages.

The advancement of time gems can possibly affect different fields, with quantum processing being perhaps of the most encouraging field. Quantum PCs bridle the standards of quantum mechanics to perform computations at speeds that outperform traditional PCs.

Time gems could assume a part in making steady and vigorous quantum bits (qubits), the essential units of quantum data.

The solidness of time precious stones emerges from the inherent idea of their time-keeping properties. The occasional swaying between various states in a period gem offers a dependable and unsurprising component for encoding and controlling quantum data. This could address a portion of the difficulties related with the delicacy of quantum states in traditional quantum figuring frameworks.

Besides, time precious stones have suggestions for how we might interpret the underpinnings of thermodynamics. The second law of thermodynamics directs that entropy, a proportion of confusion, will in general increment over the long run in disengaged frameworks. Time precious stones, notwithstanding, challenge this idea by showing occasional way of behaving without an expansion in entropy. This evident infringement of the subsequent regulation has started discussions and conversations about the idea of time precious stones and their relationship to thermodynamic standards.

The investigation of time precious stones has additionally prompted the investigation of novel quantum stages and colorful peculiarities. For example, analysts have conjectured the presence of higher-layered time precious stones, where the intermittent motions stretch out into extra aspects past the customary three spatial aspects and one worldly aspect. The quest for these extraordinary stages opens up new boondocks in the journey to figure out the principal idea of issue.

The rise of time precious stones has not been without discussion and discussion inside established researchers. A few physicists contend that the peculiarities saw in time gem examinations can be made sense of through additional customary means, like the presence of blemishes in the trial arrangement. Others battle that the interesting properties of time gems, including their soundness and particular quantum marks, challenge simple clarifications inside existing systems.

The continuous investigation of time gems isn't just a logical undertaking yet in addition a philosophical one, bringing up issues about the idea of time, evenness, and the crucial design of the universe. The breaking of time balance difficulties our natural comprehension of the bolt of time and prompts a reconsideration of the philosophical underpinnings of old style and quantum physical science.

In the mission to comprehend time precious stones, specialists are pushing the limits of trial methods and hypothetical structures. The interdisciplinary idea of this investigation includes ability from dense matter physical science, quantum data science, and thermodynamics. The coordinated effort among experimentalists and scholars is fundamental for unwinding the secrets of time precious stones and outfitting their true capacity for mechanical applications.

As the field of time precious stones keeps on advancing, specialists are investigating new roads for making and controlling these one of a kind periods of issue. The advancement of procedures to design time gems with explicit properties could make ready for useful applications in quantum advances and then some. The exchange among hypothesis and trial and error stays a main thrust in this state of the art field of exploration.

The philosophical ramifications of time precious stones reach out past the research center and into more extensive inquiries concerning the idea of time itself. The breaking of time evenness challenges our assumptions about the straight movement of time and raises the chance of elective fleeting designs in the texture of the real world. This has suggestions not just for how we might interpret the quantum world yet additionally for our more extensive philosophical examinations about the idea of the universe.

All in all, time gems address a progressive idea in the domain of quantum physical science, testing how we might interpret balance and time. The breaking of both spatial and fleeting balances in these extraordinary periods of issue opens up additional opportunities for mechanical applications, especially in the field of quantum figuring.

The acknowledgment of time precious stones in the lab has affirmed their reality as well as started banters about the central idea of time and thermodynamics.

As specialists dig further into the secrets of time gems, they are pushing the limits of how we might interpret quantum frameworks and the idea of time itself. The interdisciplinary idea of this investigation features the interconnectedness of various parts of physical science, from consolidated matter physical science to quantum data science. The continuous mission to disentangle the insider facts of time precious stones welcomes us to reexamine our suspicions about the idea of the real world and leave on an excursion into the outskirts of quantum physical science.

4.3 Quantum Tunneling and Time Travel Paradoxes

Quantum burrowing, a peculiarity established in the standards of quantum mechanics, challenges our old style instincts about obstructions and makes the way for captivating conceivable outcomes, including conversations of time travel mysteries. This peculiarity, first presented by George Gamow during the 1920s, includes the entrance of a molecule through a traditionally taboo energy obstruction. Quantum burrowing isn't just a central part of quantum mechanics yet in addition a subject that sparkles conversations about the idea of time, causality, and the potential for navigating worldly scenes.

At the core of quantum burrowing is the wave-molecule duality intrinsic in quantum frameworks. In traditional physical science, particles are considered unmistakable substances with obvious directions.

Nonetheless, quantum mechanics presents the idea that particles, like electrons, additionally show wave-like properties. This duality implies that particles can be depicted by wave works that address the likelihood appropriations of their positions.

At the point when a quantum molecule experiences an energy boundary, traditionally, it wouldn't have sufficient energy to outperform the hindrance. In any case, because of the wave-like nature of particles, there is a non-zero likelihood that the molecule's wave capability stretches out into the prohibited locale past the obstruction. This likelihood takes into account the peculiarity of burrowing, where the molecule can "burrow" through the boundary and show up on the opposite side.

The likelihood sufficiency related with quantum burrowing is represented by the Schrödinger condition, a major condition in quantum mechanics. The likelihood of burrowing is dramatically subject to the thickness and level of the hindrance. While burrowing is an uncommon occasion for thick and high hindrances, it turns out to be more likely for more slender and lower obstructions.

One of the most notable instances of quantum burrowing is tracked down during the time spent atomic combination inside stars. In heavenly centers, decidedly charged nuclear cores face a Coulomb hindrance because of their common shock. Traditional material science would anticipate that these cores can't defeat this obstruction, however quantum burrowing considers a non-no likelihood of combination, empowering the energy age that supports stars.

The standards of quantum burrowing have viable applications in different advances, like the passage diode, utilized in electronic circuits, and checking burrowing microscopy, a strategy that permits researchers to picture surfaces at the nuclear level. Notwithstanding, the ramifications of quantum burrowing stretch out past innovative applications and lead us to consider the idea of time and the chance of time travel.

One road of investigation connected with quantum burrowing is the thought of time travel mysteries. Time travel, an idea frequently consigned to the domains of sci-fi, turns into a subject of serious consideration when seen from the perspective of quantum mechanics. Hypothetical conversations no time like the present travel frequently include situations where particles burrow back through time, leading to interesting conundrums.

One such time travel mystery is the "burrowing through time" situation, where a molecule burrows back in time and cooperates with its prior self. This connection could prompt different perplexing circumstances, for example, the renowned "granddad conundrum." In this situation, a person who jumps through time might actually keep their granddad from meeting their grandma, prompting a logical inconsistency where the person who jumps through time's presence becomes self-nullifying.

Hypothetical conversations of time travel Catch 22s frequently include shut timelike bends (CTCs), ways through spacetime that circle back on themselves, taking into account time travel to the past. The presence of CTCs is an outcome of specific answers for the situations of general relativity, for example, those depicting pivoting dark openings known as Kerr dark openings.

The Kerr dark opening arrangement, when considered with regards to CTCs, brings up issues about causality and the potential for time circles. Assuming an article were to follow a CTC, it could, in principle, show up back at its beginning stage in spacetime, possibly prompting circumstances where circumstances and logical results become snared in manners that oppose our traditional comprehension.

Quantum burrowing, with its probabilistic and non-deterministic nature, presents an extra layer of intricacy to conversations of time travel. The vulnerabilities inborn in quantum mechanics could impact the results of time travel situations, adding a component of unusualness to the elements of causality.

Hypothetical proposition for time travel frequently include extraordinary builds, for example, wormholes, speculative passages in spacetime that could associate far off areas or various times. Wormholes are answers for the situations of general relativity that give easy routes through existence, possibly empowering travel to the past or far off districts of the universe.

The investigation of navigable wormholes, those that could be utilized for time travel, has prompted conversations about their security and the potential for keeping up with them long enough for down to earth use. Extraordinary matter with negative energy thickness, a substance that stays speculative and has not been noticed, is many times conjured as a vital fixing to keep a safe wormhole open.

Quantum burrowing enters the domain of time travel conversations while thinking about particles, or even whole items, burrowing through a safe wormhole starting with one spacetime point then onto the next. The probabilistic idea of quantum burrowing could impact the results of such situations, leading to a large number of potential time travel directions.

Notwithstanding, the investigation of time travel with regards to quantum burrowing raises various difficulties and mysteries. The vulnerability rule, a principal fundamental of quantum mechanics, sets that the more unequivocally the place of a molecule is known, the less definitively its force can be known, as well as the other way around. This vulnerability acquaints restrictions with the accuracy with which the underlying states of a time travel situation can be indicated.

Quantum burrowing additionally presents the test of impedance impacts, where the superposition of various quantum states can prompt helpful or disastrous obstruction. With regards to time travel, impedance impacts could impact the probabilities of various results, possibly prompting situations where certain ways through spacetime become pretty much plausible.

The investigation of time travel mysteries and quantum burrowing isn't just a practice in hypothetical physical science; it likewise dives into the way of thinking of time and the idea of causality. Inquiries regarding unrestrained choice, determinism, and the flexibility of the past arise as we think about the possible outcomes of time travel situations.

The Novikov self-consistency rule, proposed by Russian physicist Igor Novikov, recommends that the laws of physical science plot to forestall the event of conundrums in a self-steady way. As per this standard, any move made by a person who jumps through time in the past would be compelled by the laws of physical science to guarantee that it doesn't prompt an inconsistency.

The idea of shut timelike bends and the potential for time circles, while interesting, has prompted banters about the idea of causality and the chance of changing the past. The possibility that the past is fixed and unalterable, in some cases alluded to as the "block universe" model, suggests that any moves made by a person who goes back and forth through time in the past were in every case part of the reliable unfurling of history.

Time travel conversations additionally converge with the idea of perception in quantum mechanics. The job of the eyewitness in imploding the wave capability, a critical part of quantum estimation, prompts inquiries regarding how perceptions made during a time travel situation would impact the results. The demonstration of noticing a quantum framework could present intricacies and vulnerabilities that further confuse the elements of time travel.

The investigation of time travel Catch 22s and quantum burrowing isn't restricted to hypothetical conversations however reaches out to tests that test the underpinnings of quantum mechanics. Quantum ensnarement, one more quantum peculiarity where particles become associated in manners that challenge old style instincts, has been utilized to concentrate on the expected impact of future estimations on previous occasions.

Tests testing the infringement of Chime disparities, which give limits on the relationships permitted by traditional physical science, have looked to investigate the idea of causality and the chance of retroactive impacts. While these analyses don't include time travel as such, they dig into the interconnectedness of quantum frameworks and the unpretentious manners by which estimations in the present can influence the results of past estimations.

The connection between quantum mechanics and the idea of time keeps on being a subject of dynamic exploration and philosophical request. The journey to comprehend the expected ramifications of quantum burrowing on time travel situations challenges our traditional instincts and prompts us to investigate the limits of both quantum mechanics and our comprehension of the transient scene.

As scientists keep on digging into the secrets of quantum mechanics and the idea of time, the crossing point of these two domains stays a ripe ground for investigation. The philosophical ramifications of time travel situations brief us to reevaluate how we might interpret causality, through and through freedom, and the crucial design of the universe. The continuous mission to unwind the secrets of quantum burrowing and time travel oddities welcomes us to travel into the outskirts of both hypothetical physical science and the idea of reality itself.

4.4 Time's Arrow in the Quantum Realm

The bolt of time, the unidirectional stream from past to present to future, is a key part of our regular experience. However, diving into the minute domain of quantum mechanics acquaints interesting difficulties with our old style instincts finally. The idea of time's bolt in the quantum domain, where particles are dependent upon the probabilistic and reversible laws of quantum mechanics, brings up significant issues about the crucial lopsidedness among past and future.

Traditional physical science, in view of deterministic conditions of movement, suggests that the regulations administering the way of behaving of particles are time-symmetric. At the end of the day, if one somehow managed to switch the speeds and places of all particles in a traditional framework, the subsequent elements would develop in reverse in time definitively as the forward elements unfurl. Be that as it may, the plainly visible world we notice shows a reasonable inclination for specific fleeting headings, typified constantly law of thermodynamics.

The second law of thermodynamics expresses that the entropy, a proportion of the issue or haphazardness of a framework, will in general increment after some time in a disengaged framework. This prompts the natural perception that cycles in the plainly visible world are irreversible - a wrecked glass doesn't precipitously reassemble, and an ice 3D square left in a warm room softens as opposed to freezing further. The expansion in entropy is related with the directionality of time, giving a bolt that focuses from the past, where entropy was lower, to the future, where it will in general increment.

Conversely, the tiny laws of material science, especially those depicted by quantum mechanics, are time-reversible. The Schrödinger condition, a basic condition in quantum mechanics, oversees the development of quantum states and is symmetric regarding time inversion. As per the Schrödinger condition, the development of a quantum framework is similarly substantial whether time runs forward or in reverse.

The dumbfounding conjunction of time-symmetric tiny regulations and time-deviated plainly visible perceptions frames the quintessence of the bolt of time issue in physical science. Scientists have wrestled with understanding how the naturally visible bolt of time rises out of the infinitesimal, time-symmetric laws of quantum mechanics. A few methodologies and hypothetical structures have been proposed to address this problem.

One road of investigation includes the idea of quantum decoherence. Decoherence alludes to the cycle by which a quantum framework collaborating with its current circumstance loses cognizance and becomes snared with the natural levels of opportunity. This snare with the climate prompts the concealment of impedance between various quantum states, really "picking" a favored premise of states.

The development of a favored premise due to decoherence gives a characteristic directionality in time. When a framework becomes ensnared with its current circumstance, the trap commonly increments over the long haul, prompting an irreversible expansion in entropy. This association among decoherence and the bolt of time is a convincing clarification for the evident irreversibility saw in perceptible frameworks.

One more way to deal with understanding the bolt of time includes the investigation of quantum estimations. The course of estimation in quantum mechanics presents an unmistakable fleeting deviation. As indicated by the von Neumann-Wigner translation of quantum mechanics, the demonstration of estimation implodes the wave capability of a framework, deciding a positive result. This breakdown is intrinsically irreversible and presents a bolt of time related with the estimation interaction.

Notwithstanding, the subject of why estimations in the quantum domain display a favored course stays a subject of progressing exploration and discussion. A few translations, like the many-universes understanding, recommend that all potential results of an estimation happen in various parts of the quantum multiverse. The apparent bolt of time emerges from the way that our emotional experience is bound to a solitary branch, and the expanding itself is irreversible.

The idea of quantum ensnarement, where particles become connected in manners that challenge old style instincts, likewise assumes a part in conversations about the bolt of time. Entrapment presents a type of non-neighborhood connection between's particles, where the condition of one molecule is personally attached to the condition of another, regardless of whether they are generally isolated in space. The quantum entrapment process itself is time-symmetric, yet when joined with different factors, for example, decoherence, it can add to the rise of a bolt of time.

The connection between quantum mechanics and gravity, as depicted by the structure of quantum gravity, is one more region where the bolt of time issue is investigated.

Hypothetical advancements in this field, for example, the Wheeler-DeWitt condition, expect to bind together

quantum mechanics and general relativity. The transaction between quantum states and the math of spacetime with regards to quantum gravity might actually reveal insight into the beginning of the bolt of time.

The cosmological bolt of time, related with the extension of the universe, is one more feature of the bolt of time issue. The noticed extension of the universe suggests a condition of lower entropy previously, as worlds and different designs were all the more thickly stuffed. The association between the cosmological bolt of time and the minute laws of quantum mechanics stays an open inquiry, and specialists try to comprehend how the bolt of time arises with regards to the whole universe.

The idea of time's bolt in the quantum domain has suggestions for how we might interpret major material science as well as for more extensive philosophical contemplations. The idea of time, causality, and the connection between the minuscule and plainly visible sizes of the universe become focal topics in these considerations.

The philosophical ramifications of time's bolt in the quantum domain stretch out to inquiries concerning determinism and unrestrained choice. Assuming the minuscule laws of physical science are innately deterministic and time-symmetric, how does the clear indeterminacy and directionality saw in naturally visible frameworks connect with our view of through and through freedom and organization?

The interchange between the bolt of time and cognizance likewise enters the philosophical talk. The abstract insight of time, the feeling of the past affecting the present and the expectation representing things to come, brings up issues about the job of cognizance in the bolt of time issue. A few hypotheses recommend that cognizance itself assumes an essential part in the rise of time's bolt, proposing an association between the idea of cognizance and the basic design of the universe.

The journey to comprehend time's bolt in the quantum domain is a diverse undertaking that includes the crossing point of material science, reasoning, and cosmology. The bolt of time issue prompts scientists to investigate the associations between quantum mechanics, thermodynamics, and gravity, looking for a far reaching system that can accommodate the tiny and naturally visible parts of the universe.

Chapter 5

The Human Perception of Time

The human impression of time is a complicated and complex peculiarity that has captivated logicians, researchers, and scholars all through the ages. Time is a crucial part of our reality, impacting our regular routines, molding our recollections, and outlining our encounters. However, regardless of its inescapability, the idea of time stays slippery, and our comprehension of it keeps on advancing.

One of the most striking parts of human impression of time is its subjectivity. Time appears to elapse at changed rates relying upon different elements, like age, mind-set, and action. The saying "time passes quickly while no doubt about it" catches this emotional experience well. At the point when taken part in charming or engrossing exercises, time seems to accelerate, and hours can get away inconspicuous. Then again, snapshots of weariness or expectation can cause time to appear to delay perpetually.

This subjectivity isn't restricted to individual encounters; it reaches out to social and cultural settings also. Various societies might have particular transient directions, influencing how people inside those societies see and worth time. A few social orders focus on reliability and view time as a significant asset to be overseen proficiently, while others might take on a more adaptable and loosened up disposition toward worldly issues.

Worldly discernment is additionally affected by mental cycles, like consideration and memory. The "time passes quickly" peculiarity is connected to the idea of time discernment, where consideration regarding the progression of time influences its apparent span. At the point when people are completely fascinated in an action, they might forget about time, prompting a misshaped feeling of its entry.

Memory, as well, assumes an essential part in forming our view of time. The manner in which we recollect occasions can altogether impact how we see the term of time spans. For instance, a time of extraordinary and noteworthy encounters might be reflectively seen as more limited than a tedious or unremarkable one. This peculiarity is known as the "time passes quickly while you're not kidding" predisposition in review time assessment.

The connection among time and memory is additionally featured by the idea of "time enlargement" in snapshots of serious pressure or risk. A few people report encountering a dialed back view of time during dangerous circumstances, permitting them to handle data all the more rapidly and pursue split-subsequent options. This emotional contortion of time is believed to be a consequence of elevated excitement and expanded consideration, prompting a more point by point and striking memory of the occasion.

Notwithstanding emotional encounters, the human view of time is profoundly interwoven with social and cultural designs. The creation of schedules and timekeepers, for example, plays had a significant impact in normalizing and coordinating worldly encounters. These instruments work with pragmatic parts of day to day existence as well as shape the manner in which people conceptualize and explore the progression of time.

Schedules, with their cycles and divisions, give a system to sorting out occasions and exercises over longer periods. The reception of explicit schedule frameworks reflects social and strict qualities, impacting how social orders structure time and celebrate repeating occasions. Whether in light of lunar cycles, sun powered years, or other divine peculiarities, schedules act as social relics that shape the aggregate view of time.

Tickers, with their estimation of hours, minutes, and seconds, further refine the human experience of time. The accuracy and normalization presented by tickers have broad results, from organizing social and monetary exercises to managing day to day schedules.

The advancement of exact timekeeping gadgets, like mechanical clocks and later nuclear clocks, has expanded the accuracy of time estimation as well as prompted the foundation of worldwide time guidelines.

In spite of the objective estimation of time given by tickers, the emotional experience of time endures, and the transaction among level headed and abstract time is a captivating area of study. Psychophysical tests have investigated the connection between actual time stretches and saw term, uncovering that variables like consideration, excitement, and oddity can impact abstract time insight.

Transient deceptions, where people misconstrue the term of occasions, further highlight the intricacy of time discernment. For instance, the "time request mistake" happens when the apparent request of occasions varies from their genuine succession. These transient mutilations feature the job of mental cycles in molding how we might interpret time and challenge the thought of a direct and uniform insight of fleeting occasions.

The investigation of time discernment reaches out past the individual and social levels to include more extensive cultural peculiarities. Verifiable and land time scales give structures to grasping the advancement of civilizations and the actual Earth. The profound season of geographical cycles, spreading over millions and billions of years, offers a point of view that rises above individual lifetimes and mankind's set of experiences.

In the domain of material science, the idea of time takes on new aspects. The hypothesis of relativity, planned by Albert Einstein, upset how we might interpret time as a family member and dynamic substance. As indicated by relativity, time is certainly not an outright and invariant measure yet is impacted by the movement and gravitational fields of spectators. This knowledge challenges our natural ideas of time as an all inclusive steady and uncovers its complexities at the enormous level.

The association among reality, known as spacetime, is a central idea in the hypothesis of relativity. This converging of worldly and spatial aspects underlines the interconnected idea of the universe, where situation unfurl in a continuum that incorporates both fleeting and spatial directions. The bend of spacetime around enormous articles, like planets and stars, further delineates the powerful interaction among gravity and the texture of the universe.

Quantum mechanics, the part of physical science that portrays the way of behaving of particles at the littlest scales, acquaints extra layers of intricacy with the comprehension of time. In quantum hypothesis, the idea of superposition permits particles to exist in various states at the same time, testing the customary thoughts of circumstances and logical results. The immortal idea of specific quantum states difficulties our instinctive comprehension of a straight and irreversible progression of time.

The bolt of time, addressing the lopsidedness among past and future, is an idea profoundly imbued we would say. Entropy, a proportion of confusion in a framework, is frequently connected with the bolt of time. The second law of thermodynamics expresses that the entropy of a shut framework will in general increment over the long haul, prompting a directionality in actual cycles. This rising problem is connected to our abstract view of time's irreversible stream from past to future.

The investigation of time with regards to material science brings up significant issues about the idea of the real world and the impediments of human insight. The mission to comprehend a definitive nature of time has prompted speculative hypotheses, for example, the chance of time circles, equal universes, and the idea of a "block universe" where past, present, and future coincide as changeless elements.

Philosophical investigations into the idea of time date back to old civic establishments, with scholars like Heraclitus contemplating the idea of unending change and Parmenides stating the permanence of an immortal reality. Over the course of reasoning, the idea of time has been a focal topic, with different ways of thinking offering dissimilar points of view.

Immanuel Kant, in his "Scrutinize of Unadulterated Explanation," contended that time is an essential part of human discernment, deduced and important for coordinating tactile encounters. As indicated by Kant, time is definitely not an outside element yet an emotional system through which people structure how they might interpret the world. This supernatural perspective on time features its close association with human awareness.

Existentialist scholars, including Jean-Paul Sartre, investigated the existential ramifications of human fleetingness. Sartre's idea of "dishonesty" underlines the propensity of people to avoid their opportunity and obligation by capitulating to cultural assumptions and transient limitations. The attention to one's mortality and the restricted idea of individual presence add to the existential anxiety related with the human experience of time.

Interestingly, Eastern ways of thinking, for example, Buddhism, offer points of view on time that vary from Western ideas. The idea of fleetingness (anicca) in Buddhism underlines the transient and consistently changing nature, everything being equal. According to this point of view, time is certainly not a straight movement yet a ceaseless pattern of birth, rot, and resurrection, featuring the interconnectedness of all presence.

Writing and craftsmanship give roads to investigating the abstract and social elements of time. Authors and specialists frequently wrestle with the subtle idea of time, utilizing account strategies, imagery, and visual portrayals to convey the smoothness and intricacy of transient encounters. Marcel Proust's stupendous work "Looking for Lost Time" dives into the interaction between memory, time, and character, catching the complexities of human cognizance.

The visual expressions, as well, have drawn in with the subject of time in different ways. The surrealist development, for instance, looked to rise above traditional thoughts of time and reality through illusory and fantastical symbolism. Salvador Dalí's notorious painting "The Diligence of Memory," highlighting softening clocks, has turned into a significant portrayal of the liquid and abstract nature of time.

Music, with its worldly aspect, offers an extraordinary road for communicating the human experience of time. Writers use beat, rhythm, and design to make worldly scenes that summon feelings and shape the audience's impression of span. The idea of melodic time difficulties the linearity of clock time, taking into consideration a more liquid and expressive experience.

Mechanical headways in the cutting edge period have additionally changed the human experience of time. The speed increase of correspondence, transportation, and data trade has compacted transient distances and adjusted the speed of day to day existence. The quickness of computerized correspondence and the steady progression of data add to a feeling of time pressure, where people want to stay aware of a steadily speeding up present.

The appearance of the web and virtual entertainment has acquainted new aspects with fleeting encounters. The immediate dispersal of data across the globe breakdowns fleeting and spatial limits, making a feeling of synchronization and interconnectedness. Web-based entertainment stages, with their ongoing updates and consistent network, add to an impression of time as a ceaseless and solid stream of online collaborations.

At the same time, the advanced age has led to new difficulties in overseeing consideration and supporting concentration. The consistent assault of boosts and the appeal of moment satisfaction in the advanced domain can upset conventional rhythms of work and recreation, impacting the abstract insight of time. The mixing of work and individual life in the web-based climate further confounds the depiction of worldly limits.

The connection among innovation and time is additionally obvious in the improvement of man-made reasoning and the investigation of potential prospects where machines have their own worldly cognizance. Speculative conversations about the peculiarity, a theoretical place where man-made reasoning outperforms human knowledge, bring up issues about the idea of machine time and its suggestions for the human experience.

The idea of time travel, a well known topic in sci-fi, dazzles the creative mind and difficulties our instincts about causality and the strength of the past. While time travel stays a speculative thought, the investigation of its speculative situations in writing and film gives an imaginative space to considering the idea of time and its likely intricacies.

5.1 Psychological Time: Subjective Experiences and Mental Time Travel

Mental time, a feature of human cognizance, includes the emotional encounters and mental cycles through which people see, decipher, and recall time. Dissimilar to the objective estimation of time given by clocks and schedules, mental time digs into the perplexing manners by which the brain builds and explores the transient scene. This investigation drives us to the idea of mental time travel, a mental capacity that permits people to extend themselves into the past and future, molding their discernments and choices.

The abstract insight of time is profoundly entwined with consideration, feeling, and memory. Our view of time can fluctuate in light of the degree of consideration we commit to various exercises. At the point when completely participated in an undertaking or submerged in a pleasurable encounter, time seems to elapse rapidly — an impact usually portrayed as "time passes quickly while you're not kidding." On the other hand, times of fatigue or dreariness can cause time to appear to delay, featuring the flexibility of transient discernment.

Close to home states likewise assume an essential part in molding how we experience time. High excitement feelings, like trepidation or energy, can prompt a bending of time insight. In snapshots of extreme pressure, people might report a feeling of time dialing back — a peculiarity frequently connected with the striking encoding of recollections during basic occasions. The close to home valence of an encounter can impact whether time is recognized as passing or delayed, adding to the rich woven artwork of our fleeting recollections.

Memory, a critical part of mental time, goes about as both a recorder and shaper of our worldly encounters. The manner in which we recall occasions can fundamentally affect our view of the length of time stretches. The "memory knock," a peculiarity where people review a lopsided number of personal recollections from late immaturity and early adulthood, mirrors the impact of memory on our feeling of time. The importance and profound remarkable quality of occasions during this period add to a packed and clear memory of the past.

The connection among time and memory is additionally exemplified by the idea of "forthcoming memory," which includes making sure to play out an arranged activity later on. Effectively executing expectations depends on the capacity to intellectually project oneself forward in time and expect future occasions. Disappointments in forthcoming memory, for example, neglecting to go to a planned arrangement, highlight the difficulties innate in the mental cycles that span the present and what's in store.

Mental time travel, a mental capacity interesting to people, permits people to intellectually return to the past and mimic potential fates. This limit with regards to mental time travel is complicatedly connected to the improvement of an identity and individual character.

The capacity to consider one's previous encounters and imagine future situations is fundamental to the development of a durable and constant identity over the long run.

Roundabout memory, a type of long haul memory that stores data about unambiguous occasions and encounters, is a critical part of mental time travel. The reactivation of wordy recollections includes intellectually reproducing the subtleties of previous occasions, permitting people to drench themselves in the sights, sounds, and feelings related with those minutes. The extravagance of rambling memory adds to the striking quality of mental time travel, empowering people to return to and remember previous encounters in their imagination.

The forward-looking part of mental time travel includes prospection, the capacity to expect and anticipate future occasions. Prospection isn't restricted to basic objective setting; it stretches out to the reenactment of different situations and the thought of expected results. The useful idea of mental time travel permits people to imagine elective fates, make arrangements, and make moves in the present with an eye toward their expected results.

The prefrontal cortex, a district of the cerebrum related with chief capabilities and navigation, assumes an essential part in mental time travel. Neuroscientific studies have recognized brain networks associated with both the recovery of roundabout recollections and the reenactment of future situations. The covering brain processes hidden recalling the past and envisioning the future feature the interconnectedness of memory and prospection in the development of mental time.

While mental time travel gives a significant versatile capability, permitting people to gain from an earlier time and plan for the future, it isn't resistant to predispositions and contortions. The peculiarity of "knowing the past inclination," where occasions appear to be more unsurprising and expected after they have happened, mirrors the reconstructive idea of memory. Looking back, people might see previous occasions as having been more predictable than they really were, possibly affecting their evaluation of their own critical thinking skills.

Firmly related is the "estimating blunder," an inclination to misconstrue the force and span of future profound states. While imagining future situations, people frequently battle to foresee the profound effect of occasions, prompting mistakes in emotional guaging precisely. The inconsistency among anticipated and genuine close to home responses adds to the difficulties of independent direction and objective pursuit.

The crossing point of mental time and navigation turns out to be especially notable with regards to transient limiting — the inclination to appoint lower worth to deferred rewards contrasted with quick rewards. Fleeting limiting mirrors the intrinsic pressure between momentary cravings and long haul objectives. People might battle to settle on decisions that line up with their future prosperity, capitulating to the appeal of quick delight.

The experience of time in the computerized age presents new difficulties and open doors for mental time. The consistent availability worked with by cell phones and the web has changed the speed of data trade and correspondence. The instantaneousness of computerized cooperations has suggestions for consideration the board and the nature of transient encounters. The consistent flood of warnings and the draw of moment delight add to a feeling of time pressure, where people feel a sense of urgency to answer rapidly to the requests of the web-based climate.

Virtual entertainment, with its ongoing updates and organized portrayals of life, shapes the manner in which people see their own day to day routines and the existences of others. The ceaseless stream of data via web-based entertainment stages makes a feeling of concurrence, where situation unfurl in a never-ending present. The organized idea of online substance further foggy spots the line among the real world and portrayal, affecting the development of individual accounts and fleeting encounters.

The idea of "FOMO" (Feeling of dread toward Passing up a major opportunity) embodies the mental effect of web-based entertainment on transient encounters. People might feel compelled to take part in exercises or occasions, driven by the apprehension about passing up shared encounters. This apprehension about passing up a major opportunity adds to an elevated familiarity with time elapsing and the longing to remain associated progressively, even to the detriment of present-second encounters.

The computerized domain likewise presents the peculiarity of "time moving," where people take part in nonconcurrent correspondence and media utilization. The capacity to get to data, diversion, and social associations whenever disturbs conventional fleeting limits. Web-based features, on-request happy, and nonconcurrent correspondence stages give people the adaptability to consume and make content external the requirements of straight time.

The effect of innovation on mental time reaches out to the domain of work and efficiency. Remote work, worked with by computerized specialized instruments, challenges customary ideas of a decent working day and available time. The obscuring of limits among work and individual life in the web-based climate brings up issues about the nature of relaxation time and the capacity to detach from business related requests.

The experience of time in the computerized age isn't uniform, and people explore these difficulties in assorted ways. Some might embrace the valuable open doors for adaptability and independence managed the cost of by advanced innovations, while others wrestle with the requests of consistent availability and data over-burden. The mental ramifications of these mechanical movements highlight the requirement for a nuanced comprehension of how people see and deal with their transient encounters in the contemporary world.

Social and cultural elements contribute altogether to the forming of mental time. Various societies might show particular fleeting directions, affecting how people inside those societies see and worth time. The idea of "polychronic" and "monochronic" societies, presented by anthropologist Edward T. Lobby, features social varieties in the way to deal with time. Polychronic societies underline the liquid and adaptable nature of time, taking into consideration various exercises to all the while happen. Conversely, monochronic societies focus on successive and straight ways to deal with time, esteeming dependability and adherence to plans.

Social mentalities toward time likewise manifest in customs, functions, and festivities that mark the progression of time. The repeating idea of comprehensive developments, like occasions and celebrations, mirrors a common feeling of fleeting coherence and cadence. These social practices add to a common perspective of time and make a feeling of aggregate personality moored in worldly encounters.

5.2 Cultural Variations in Time Perception

Social varieties in time discernment enlighten the assorted manners by which various social orders consider, insight, and worth the progression of time. Time, while generally experienced, is definitely not a solid idea; rather, it is molded and deciphered by social structures, verifiable settings, and cultural standards. Understanding how societies see and coordinate time gives important bits of knowledge into the perplexing transaction between worldly encounters and social designs.

The idea of time direction is a fundamental part of social varieties in time discernment. Anthropologist Edward T. Lobby presented the qualification among monochronic and polychronic societies to catch various ways to deal with time. Monochronic societies, pervasive in numerous Western social orders, underscore a straight and successive perspective on time. In these societies, reliability, adherence to plans, and proficient using time effectively are exceptionally esteemed. Time is in many cases seen as need might arise to be distributed and used actually.

Then again, polychronic societies embrace a more liquid and adaptable way to deal with time. These societies focus on relational connections and the fulfillment of errands over unbending adherence to plans. Synchronous commitment to numerous exercises is acknowledged and anticipated, mirroring a mutual direction toward time. Polychronic societies frequently display a higher capacity to bear uncertainty and a readiness to adjust plans in view of social communications and relevant variables.

The social components of monochrony and polychrony manifest in ordinary practices and normal practices. For instance, conferences in monochronic societies regularly follow an organized plan, and members are supposed to show up and close the gathering on time.

Conversely, polychronic societies might put a more prominent accentuation on the social parts of gatherings, taking into consideration more adaptable timetables and unconstrained conversations.

Social varieties in time discernment are not bound to the fleeting association of day to day exercises however reach out to more extensive cultural designs. The idea of "occasion time" and "clock time," presented by humanist Eviatar Zerubavel, outlines the different transient structures that shape social communications. In social orders described by occasion time, the planning of not entirely settled by the event of explicit occasions or achievements as opposed to by the clock. This transient direction is many times seen in agrarian social orders or conventional networks where farming cycles and shared occasions impact worldly rhythms.

Then again, clock time, predominant in industrialized and urbanized social orders, underscores exact booking and adherence to normalized time units. Industrialization, the ascent of the advanced week's worth of work, and the presentation of normalized time regions have added to the predominance of clock time in numerous contemporary societies. The synchronization of exercises in view of clock time works with coordination in complex social frameworks yet in addition presents a degree of regimentation and time pressure.

Strict and philosophical customs further add to social varieties in time discernment. The repetitive origination of time, established in numerous Eastern ways of thinking and religions, appears differently in relation to the direct view predominant in Judeo-Christian customs. In Hinduism and Buddhism, for instance, time is viewed as repeating, portrayed by the persistent pattern of creation, conservation, and disintegration. The recurrent idea of time lines up with the idea of rebirth, where people go through a progression of births and passings.

Interestingly, the Judeo-Christian practice places a straight perspective on time, set apart by a conclusive start and end. The direct movement of time is implanted in strict accounts, like the creation story in Beginning and the prophetically calamitous dreams in the Book of Disclosure. This straight system shapes social understandings of progress, history, and the unfurling of heavenly plans.

The festival of New Year's Day gives an illustrative illustration of social varieties in time discernment. While the progress to another year is generally perceived, various societies might credit unmistakable implications and ceremonies to this worldly milestone. Western societies frequently view the new year as a chance for goals, self-improvement, and the hug of additional opportunities. Conversely, different societies might stress coherence, familial associations, and the respecting of customs during the progress to another year.

Social varieties in time discernment are not static; they develop in light of cultural changes, globalization, and mechanical progressions. The reception of a globalized fleeting structure, set apart by normalized time regions and facilitated worldwide occasions, has impacted the transient encounters of people across different societies. The globalization of business, correspondence, and amusement has prompted the dispersion of Western thoughts of using time effectively and proficiency to various areas of the planet.

Mechanical progressions, especially in correspondence and transportation, have additionally sped up the speed of social trade and affected worldly encounters. The immediate idea of computerized correspondence and the capacity to cross significant stretches in a brief period add to a feeling of fleeting pressure. Societies might wind up exploring the pressure between customary fleeting directions and the requests of an interconnected, speedy world.

Social varieties in time discernment are likewise reflected in etymological articulations and illustrations connected with time. The Sapir-Whorf speculation proposes that language reflects as well as molds suspected. In the domain of time discernment, phonetic contrasts add to different social originations of worldly encounters. For instance, a few dialects might utilize spatial similitudes to depict time, for example, alluding to the past as behind and the future as ahead. Different dialects might utilize cyclic illustrations, underscoring the common idea of time.

The Aymara language spoken in the Andes locale gives a special delineation of how phonetic designs can impact transient discernment. In Aymara, the past is etymologically portrayed as lying in front, while what's in store is enunciated as behind. This semantic reversal challenges the ordinary spatial similitudes utilized in numerous Western dialects and highlights the variety of social viewpoints on time.

Social varieties in time discernment likewise meet with the idea of "transient limiting," the propensity to relegate various qualities to present and future results. In societies that focus on quick friendly connections and public congruity, people might display lower worldly limiting, esteeming potential compensations all the more comparatively to prompt prizes. Conversely, societies that accentuate individual accomplishment and future-arranged objectives might display higher fleeting limiting, focusing on transient additions over postponed rewards.

The effect of social varieties in time discernment stretches out to the domain of training and mental turn of events. Multifaceted examinations have investigated how social contrasts in fleeting directions impact mental cycles, including memory, consideration, and critical thinking. For instance, research recommends that people from polychronic societies might exhibit a more noteworthy capacity to bear interferences and performing various tasks, mirroring the versatile mental cycles related with their fleeting direction.

In multicultural social orders, people frequently explore between social structures, adjusting their fleeting ways of behaving and assumptions in view of the unique situation. This intercultural smoothness presents the idea of "biculturalism" or "multiculturalism" in time discernment, where people draw according to various social viewpoints in their worldly encounters. The discussion of social varieties in time discernment turns out to be especially remarkable in different work environments, instructive settings, and relational connections.

Social clinicians and anthropologists stress the significance of embracing a "emic" viewpoint, taking into account the inner social implications and values related with time inside a particular social setting. Emic approaches perceive that assorted societies might have remarkable fleeting standards, ceremonies, and accounts that are fundamental to the social texture of those social orders. The enthusiasm for social variety in time discernment requires a nuanced understanding that goes past oversimplified correlations and recognizes the extravagance of social implications related with time.

5.3 Aging and the Perception of Time

Maturing is an all inclusive and inescapable part of the human experience, set apart by an intricate interchange of physiological, mental, and social changes. One fascinating component of maturing is its significant effect on the view of time. As people age, their abstract insight of time goes through movements and changes that impact how they consider the past, draw in with the present, and expect what's to come.

One of the observable parts of maturing and time discernment is the peculiarity frequently alluded to as "time pressure." As people age, they frequently report a feeling that time is elapsing more rapidly than it did in their previous years. This discernment might be credited to a few elements, including routine and commonality. As day to day schedules become more imbued and routine, the uniqueness of individual days might reduce, adding to a sensation of time elapsing quickly.

The differentiation between the apparent speed of time in youth versus later adulthood is a typical episodic perception. Kids frequently experience time as broad and apparently unending, with every day loaded up with novel encounters and learning. Conversely, as people age and sink into laid out schedules, the days might mix together, bringing about a review sense that time has sped up.

Mental clinicians recommend that the job of memory has a urgent impact in this time pressure peculiarity. The quantity of novel and vital occasions will in general diminish with age, and accordingly, there might be less unmistakable markers in memory that assist people with checking the progression of time. The striking quality of cherished, lifelong recollections, interestingly, adds to the impression of an additional lengthy and significant past.

The mental handling speed, one more angle affected by maturing, may add to changes in time discernment. Research demonstrates that as people age, there is a general decrease in handling speed. This mental easing back can influence the manner in which people see the term of occasions and timespans. Undertakings that once appeared to be quick and easy may now demand greater investment and exertion, possibly impacting the general feeling of time elapsing.

The "time passes quickly" peculiarity, frequently connected with the inclination that time speeds up as one ages, features the subjectivity of fleeting encounters. Hypotheses connected with time insight, for example, the "corresponding hypothesis," place that our impression of time is comparative with the aggregate sum of time we have encountered. Thus, a year might feel more limited to a more established person as it addresses a more modest extent of their lifetime contrasted with a more youthful individual.

Fleeting milestones, like birthday celebrations, additionally add to the impression of time with regards to maturing. While birthday celebrations are customarily events for festivity, they can likewise act as tokens of the progression of time and the maturing system. People might participate in review assessments of their accomplishments and life achievements, impacting their view of how rapidly or gradually time has elapsed.

The mental build of "time viewpoint" gives a structure to understanding how people situate themselves corresponding to past, present, and future occasions. Zimbardo and Boyd proposed a model that sorts people in view of their prevailing time viewpoint direction: past-situated, present-arranged, or future-arranged. As people age, there might be shifts in their time point of view, impacted by formative changes, valuable encounters, and moving needs.

In early adulthood, people might display a future-situated time point of view, portrayed by an emphasis on long haul objectives, profession desires, and anticipating what's in store. As people progress through the phases of life, there might be a shift towards a more adjusted viewpoint that integrates previous encounters, present happiness, and future objectives. In more seasoned age, a past-situated time point of view might turn out to be more noticeable, with an expanded spotlight all things considered, memory, and the tradition of one's life.

The mental peculiarity of "time skylines" likewise assumes a part in how people approach time as they age. Time skylines allude to the apparent fleeting distance between the present and future occasions. In youth, people might have longer time skylines, contemplating far off objectives and potential outcomes. Nonetheless, as people age, the consciousness of restricted time remaining may prompt more limited time skylines, with an emphasis on capitalizing on the present and short term.

Social and cultural variables cross with maturing and time insight. Various societies might have special assumptions and standards in regards to the maturing system, impacting how people see their own maturing and the progression of time. In societies that esteem intergenerational associations and the insight of elderly folks, maturing might be related with a feeling of satisfaction and commitment. Conversely, societies that underscore energy and efficiency might add to negative impression of maturing.

The idea of "worldly limiting" additionally becomes possibly the most important factor with regards to maturing. Worldly limiting alludes to the propensity to cheapen potential compensations for sure fire satisfaction. In the domain of maturing, people might wrestle with choices that include compromises between momentary delights and long haul prosperity. Wellbeing related decisions, monetary preparation, and way of life choices may be in every way affected by the harmony between prompt fulfillment and future results.

The effect of maturing on the impression of time isn't exclusively a mental peculiarity; it is profoundly interwoven with close to home encounters. The consciousness of restricted time staying in life can evoke a scope of feelings, including wistfulness, lament, and an uplifted appreciation for significant encounters. Wistfulness, specifically, includes a nostalgic yearning for the past and may turn out to be more articulated as people age, adding to a positive or self-contradicting viewpoint on the progression of time.

Research recommends that profound prosperity and life fulfillment in later years might be connected to the capacity to build a rational life story. As people participate during the time spent life survey, incorporating previous encounters into a significant story can add to a feeling of direction and satisfaction. The manner in which people decipher and approach their biographies impacts their impression of the past as well as their expectation representing things to come.

The idea of "time left throughout everyday life," rather than sequential age, turns into a critical consider molding people's mentalities toward maturing. The view of time left in life is abstract and fluctuates across people in light of their wellbeing, objectives, and individual conditions. People who see more than adequate time left might move toward maturing with a feeling of chance and proceeded with development. On the other hand, the individuals who see time as restricted may encounter a need to get a move on and an emphasis on heritage and significant associations.

The job of social connections and social help turns out to be progressively significant with regards to maturing and time insight. Forlornness and social disconnection, predominant among more established grown-ups, can influence the view of time and add to a feeling of vacancy or dullness. On the other hand, positive social collaborations, friendship, and a feeling of having a place can improve the nature of transient encounters in later life.

The job of memory in molding the impression of time in maturing is a mind boggling and complex perspective. While mental maturing might prompt decreases in specific parts of memory capability, research recommends that more established grown-ups may display an energy predisposition in memory review. This predisposition includes a memorable inclination positive occasions more distinctively than adverse occasions, adding to a rosier review perspective on life.

The effect of neurobiological changes on time discernment is one more area of investigation in maturing research. The cerebrum structures engaged with time handling, like the prefrontal cortex and the hippocampus, go through changes with age. These progressions might impact the precision of worldly decisions and the joining of transient data into the more extensive setting of encounters.

5.4 The Neuroscience of Time: How the Brain Encodes Temporal Information

The many-sided and complex nature of time discernment has long captivated researchers and analysts, prompting the investigation of the neuroscience basic how the mind encodes fleeting data. The brain systems liable for our view of time include a perplexing interaction of cerebrum locales and cycles, enveloping both the impression of outer transient occasions and the emotional experience of time.

One major part of the neuroscience of time insight includes the idea of brain timing components. These instruments are liable for estimating and addressing the length of occasions, spans, and fleeting successions. While the specific brain instruments stay an area of dynamic examination, studies recommend the contribution of appropriated brain networks that work across different cerebrum districts.

The prefrontal cortex, a locale related with leader capabilities, direction, and working memory, assumes a vital part in fleeting handling. Concentrates on utilizing neuroimaging methods, for example, utilitarian attractive reverberation imaging (fMRI), have uncovered expanded prefrontal cortex enactment during undertakings including fleeting decisions. The dorsolateral prefrontal cortex, specifically, has all the earmarks of being ensnared in undertakings requiring the assessment and examination of time spans.

The job of the cerebellum in timing and fleeting handling has additionally gathered consideration. Customarily connected with engine control, the cerebellum is currently perceived for its association in non-engine capabilities, including transient handling. Sore examinations and neuroimaging research propose that the cerebellum adds to the precise timing of engine activities and may assume a part in the view of transient spans.

The basal ganglia, a gathering of subcortical designs engaged with engine control and prize handling, is one more central member in the brain components of time discernment. Research shows that disturbances in basal ganglia capability can prompt timing debilitations, influencing the capacity to assess and repeat time stretches precisely. The complicated associations between the basal ganglia and other cortical and subcortical locales feature its job in the more extensive brain network overseeing time discernment.

The impression of worldly request, a principal part of time handling, includes the mix of tactile data across various modalities. Studies have recognized the prevalent fleeting sulcus as a locale ensnared in the handling of transient request and the combination of visual and hear-able worldly signs. The synchronization of brain action across tactile modalities is fundamental for building a rational impression of the transient grouping of occasions.

Synapse frameworks, for example, the dopaminergic framework, additionally add to the brain premise of time discernment. Dopamine, a synapse related with remuneration handling and inspiration, has been ensnared in worldly handling and span timing. Pharmacological examinations utilizing dopaminergic drugs propose a modulatory job of dopamine in fleeting decisions, with modifications in dopamine levels influencing the exactness of time discernment.

The peculiarity of "time cells" has arisen as an interesting area of examination inside the neuroscience of time insight. Like spot cells in the hippocampus that encode spatial data, time cells are neurons that display particular terminating designs connected with explicit minutes in time. Concentrates on in rodents have recognized time cells in the hippocampus and entorhinal cortex, recommending that these districts might assume a part in encoding worldly data inside the setting of verbose memory.

The hippocampus, customarily connected with spatial route and memory, gives off an impression of being engaged with the incorporation of worldly data into roundabout memory. The multifaceted interaction between the hippocampus and the prefrontal cortex is remembered to add to the development of worldly arrangements and the association of occasions inside the structure of memory. Brokenness in these brain circuits might underlie fleeting handling shortages saw in conditions like amnesia.

The peculiarity of "time traveling" further highlights the powerful idea of worldly handling in the cerebrum. Time traveling alludes to the versatile recalibration of transient insight in light of the specific situation and the consistency of occasions. Neuroimaging studies propose that the impression of time can be affected by mental and logical elements, with the mind deftly changing its fleeting appraisals in view of the data accessible.

Worldly deceptions, where people see the length or request of occasions erroneously, give extra experiences into the brain premise of time insight. The "time request mistake," for instance, happens when the apparent request of occasions varies from their genuine succession.

Brain models suggest that the transient restricting of tactile occasions depends on the synchronization of brain movement across various tangible modalities and mind areas.

The brain components fundamental the view of sub-second spans, frequently alluded to as the "millisecond timing" framework, include fast brain cycles and input circles. Research recommends the contribution of subcortical designs, for example, the basal ganglia and cerebellum, in millisecond timing. The mind boggling coordination of brain action inside these subcortical designs takes into account the exact encoding and separation of short time frame spans.

Conversely, the handling of longer time stretches, frequently named the "second timing" framework, draws in cortical and subcortical districts in a disseminated network. The inclusion of the prefrontal cortex, basal ganglia, and valuable engine region proposes that more extended time stretches require the coordination of mental cycles, engine arranging, and attentional instruments. The disseminated idea of the second timing framework features the intricacy of transient handling across various timescales.

The abstract insight of time, frequently alluded to as "mental time," includes the combination of transient data with higher-request mental capabilities. The impression of the progression of time, the attention to fleeting request, and the capacity to intellectually reenact future occasions all add to mental time. Neuroimaging studies demonstrate the commitment of the default mode organization, an organization of cerebrum locales dynamic during rest and self-referential handling, in mental time travel and the development of future situations.

Neurological problems and conditions can give significant bits of knowledge into the brain premise of time discernment. Conditions like Parkinson's sickness, described by brokenness in the basal ganglia, may prompt timing weaknesses and hardships in assessing time stretches. Investigations of patients with sores to explicit mind districts have likewise uncovered separations in worldly handling, with various cerebrum regions adding to unmistakable parts of time discernment.

The brain premise of time discernment reaches out past the handling of outside fleeting occasions to remember the mind's portrayal of the self for time. The idea of "fleeting mindfulness" includes the mix of one's past, present, and future selves into a strong and consistent feeling of character. Neuroimaging studies recommend that the default mode organization, alongside the average prefrontal cortex, may assume a focal part in the brain portrayal of oneself across time.

The connection among feelings and time insight acquaints one more layer of intricacy with the neuroscience of time. Profound occasions can regulate the view of time, with genuinely charged encounters frequently prompting changes in transient decisions.

The amygdala, a critical construction in the profound handling organization, co-operates with other fleeting handling locales to impact the emotional experience of time in sincerely remarkable circumstances.

Brain adaptability, the mind's capacity to adjust and rearrange because of encoun-ters, is a pivotal part of the neuroscience of time discernment. Natural improvement, opportunities for growth, and openness to novel boosts can shape the brain circuits engaged with time handling. Concentrates on in creature models have shown the limit with regards to brain pliancy in fleeting handling districts, featuring the unique idea of the mind's portrayal of time.

Chapter 6

Time Travel: Fact or Fiction?

Time travel has long caught the human creative mind, starting vast discussions and hypotheses about its possibility. While time travel has been a well known topic in sci-fi writing and movies, whether or not it very well may be a reality in the domain of physical science stays an enticing secret. In this investigation, we will dig into the idea of time travel, looking at the logical speculations, philosophical ramifications, and the social effect of this puzzling thought.

The thought of time as a direct movement, from past to present to future, is profoundly imbued in how we might interpret the world. Nonetheless, as physicists dove into the secrets of the universe, they started to think about how conceivable it is that time probably won't be pretty much as clear as it appears. Albert Einstein's hypothesis of relativity, planned in the mid twentieth 100 years, changed how we might interpret existence.

As per Einstein's hypothesis, time is relative and can be affected by elements like gravity and speed. The popular condition E=mc^2 showed the relationship of energy and mass, proposing that time itself could be a moldable aspect. While Einstein's situations didn't expressly embrace time travel, they made the way for the possibility that time could be controlled under specific circumstances.

One idea that arose out of the situations of relativity is time widening. Time widening happens when an item is moving for a critical portion of the speed of light or is in areas of strength for a field. In such cases, time seems to elapse all the more leisurely for the moving item or in the extraordinary gravitational field, comparative with a fixed eyewitness. This peculiarity has been tentatively affirmed through examinations including rapid particles and nuclear timekeepers put on quick planes or in solid gravitational fields.

While time enlargement is a genuine and quantifiable impact, it is fundamental to recognize time expansion as anticipated by relativity and the more fantastical thought of time travel as depicted in sci-fi. Time expansion, as depicted by Einstein's situations, doesn't take into account in reverse time travel. It only proposes that time can elapse at various rates for various onlookers in unambiguous circumstances.

The idea of voyaging in reverse in time, a staple of endless sci-fi stories, includes a large group of conundrums and difficulties. One of the most renowned Catch 22s is the "granddad mystery." Assuming time travel were conceivable, one could travel once more into the past and coincidentally keep one's grandparents from meeting, in this way forestalling one's own reality. This makes a legitimate inconsistency - on the off chance that you were rarely conceived, how is it that you could have turned back the clock in any case?

The granddad Catch 22 and comparable consistent problems have driven numerous physicists to scrutinize the attainability of in reverse time travel. Some recommend that the laws of physical science could keep such mysteries from happening, proposing the presence of "astronomical control" that shields the texture of spacetime from reckless fleeting circles. Others contend that the simple presence of these oddities might be proof that regressive time travel is innately unimaginable.

Regardless of the difficulties presented by Catch 22s, a few hypothetical physicists have investigated the numerical systems that might actually take into consideration time travel. One such idea is shut timelike bends (CTCs), proposed by physicist Kurt Gödel in 1949 and later promoted by physicist Forthcoming Tipler. CTCs are ways through spacetime that circle back on themselves, hypothetically taking into account time travel.

The chance of CTCs relies on the presence of fascinating matter with negative energy thickness, which could make the important circumstances for twisting spacetime such that licenses time circles. While the idea is numerically charming, the functional difficulties, including the speculative prerequisite for intriguing matter and the aversion of Catch 22s, make the acknowledgment of CTCs profoundly speculative.

One more hypothetical system for time travel includes navigable wormholes. Wormholes are speculative alternate routes through spacetime that could interface far off districts or even various times. Physicists Michael Morris, Kip Thorne, and Ulvi Yurtsever investigated the possibility of navigable wormholes in a 1988 paper, proposing that the making of a steady wormhole could empower time travel.

Nonetheless, safe wormholes, similar to CTCs, require extraordinary matter with negative energy thickness to keep them open. The presence and solidness of such intriguing matter remain simply speculative, and the infringement of energy conditions expected for navigable wormholes brings up critical issues about their plausibility.

While these hypothetical structures give a premise to examining the capability of time travel inside the setting of general relativity, they are a long way from being laid out logical realities. Hypothetical physical science frequently includes investigating numerical conceivable outcomes that might have actual partners in the universe. At this point, there is no exploratory proof supporting the presence of CTCs or safe wormholes.

In the domain of quantum mechanics, one more part of material science that oversees the way of behaving of particles at the littlest scales, the idea of time assumes an alternate personality. Quantum mechanics presents the idea of superposition, where particles can exist in numerous states all the while. The renowned psychological study known as Schrödinger's feline outlines this guideline, wherein a feline in a fixed box can be at the same time alive and dead until noticed.

A few physicists have investigated the possibility that quantum entrapment, a peculiarity where particles become connected in manners that challenge old style clarifications, could assume a part in time-related peculiarities. Notwithstanding, the association between quantum ensnarement and time travel stays speculative and misses the mark on strong hypothetical establishment.

As researchers wrestle with the intricacies of time inside the structures of relativity and quantum mechanics, the philosophical ramifications of time travel have likewise enraptured masterminds across disciplines. The idea of time and its relationship to human experience brings up significant issues about determinism, freedom of thought, and the actual texture of the real world.

Thinker J. M. E. McTaggart broadly ordered two perspectives on time as the "A-series" and the "B-series." The A-series includes the ideas of past, present, and future, making a dynamic and emotional experience of time. The B-series, then again, regards time as a progression of occasions requested by the relations of sooner than, concurrent with, or later than, stripping away the emotional experience of the current second.

The A-series and B-series perspectives on time add to the continuous discussion about the idea of time and its relationship to our insight. In the event that time is just a B-series, without any trace of an innate present second, it challenges our natural comprehension of the progression of time and the truth of the current second.

The idea of time travel further confuses these philosophical conversations. Assuming time is pliant and one can cross among past and future, does this imply that the past, present, and future exist all the while? Or on the other hand is there a special edge of reference that characterizes the current second?

The philosophical discussions encompassing time travel likewise address the issue of choice. In the event that what's to come is foreordained and available through time travel, it brings up issues about the independence of human decisions. Might one at any point genuinely redirect occasions, or is everything predetermined in a deterministic universe?

The convergence of material science and reasoning in regards to time travel turns out to be considerably more unpredictable while thinking about the idea of awareness. Assuming time travel were conceivable, it would challenge how we might interpret character and the steadiness of self across various fleeting places. The philosophical idea of individual character, investigated by scholars like Derek Parfit, meets with the chance of time travel, provoking inquiries concerning whether an individual could exist in different cases across time.

As conversations no time like the present travel unfurl in the domains of physical science and reasoning, the social effect of this idea has been significant. Time travel has turned into an omnipresent topic in writing, movies, and TV, catching the public's creative mind and rousing endless stories that investigate the results and conceivable outcomes of fleeting excursions.

In writing, fundamental works like H.G. Wells' "The Time Machine" established the groundwork for time travel accounts. Distributed in 1895, Wells' novella presented the idea of a machine that could ship a person through time, offering a brief look into the far off future. Wells' investigation of social and transformative topics through time travel set up for ensuing works in the class.

In the domain of sci-fi writing, writers like Isaac Asimov, Arthur C. Clarke, and Philip K. Dick further extended the account prospects of time travel. Asimov's "The Finish of Endlessness" dove into the possibility of a time-traveling association that changes verifiable occasions to improve mankind, while Clarke's "Time Odyssey" series investigated time travel inside the setting of vast secrets.

Philip K. Dick, known for his psyche bowing and reality-bending stories, utilized time travel as a topical gadget to investigate the idea of the real world and human discernment. Works like "Ubik" and "A Scanner Hazily" grandstand Dick's capacity to obscure the limits between past, present, and future, making stories that challenge ordinary ideas of time.

The effect of time travel on mainstream society is maybe most clear in the domain of film. Various movies have handled the subject of time travel, going from carefree comedies to mind-bowing spine chillers. "Back to What was to come," coordinated by Robert Zemeckis, stays a social standard, mixing humor and experience with the complexities of time travel.

In the film, young person Marty McFly turns back the clock utilizing a DeLorean vehicle outfitted with a motion capacitor, experiencing his folks in their childhood and unintentionally changing the direction of history. The outcome of "Back to What's in store" produced continuations and cemented time travel as a convincing and engaging story gadget.

Christopher Nolan's "Commencement" investigated the control of time inside the structure of dreams, testing the crowd's impression of the real world. The film's many-sided plot and enhanced visualizations added to its basic recognition and further showed the true to life capability of fleeting subjects.

TV series have likewise embraced time travel as a focal component of their stories. "Specialist Who," a long-running English sci-fi series, includes the Specialist, a Period Master from the planet Gallifrey, who goes through reality in the TARDIS (Time and Relative Aspect in Space). The show's investigation of verifiable occasions and modern situations has made it a cherished social peculiarity.

The social interest with time travel stretches out past writing and film to impact conversations in mainstream researchers. While physicists thoroughly stick to the standards of relativity and quantum mechanics in their exploration, the charm of time travel as an idea flashes interest and hypothesis even among the people who perceive its hypothetical difficulties.

Time travel has likewise penetrated well known talk, turning into a subject of interest in open discussions, digital recordings, and virtual entertainment. Novice lovers and specialists the same take part in conversations about the likely ramifications of time travel, frequently obscuring the lines between logical request and creative hypothesis.

Lately, headways in how we might interpret key material science have prompted recharged interest in the idea of time. The revelation of gravitational waves, swells in spacetime brought about by calamitous occasions in the universe, has given another apparatus to examining the texture of spacetime itself. Scientists trust that the investigation of gravitational waves might offer experiences into the idea of time and its conduct in outrageous vast circumstances.

The journey to comprehend time travel has likewise met with advancements in quantum processing. Quantum PCs, which influence the standards of quantum mechanics to perform estimations at speeds out of reach by traditional PCs, bring up issues about the expected control of time-related peculiarities at the quantum level.

While these headways add to the more extensive logical investigation of time, they miss the mark concerning giving substantial proof to the plausibility of time travel. The hole between hypothetical prospects and trial approval stays a critical test in the investigation of time and its possible control.

As our investigation of time travel attracts to a nearby, obviously the idea exists at the convergence of science, reasoning, and culture. While hypothetical structures inside material science offer numerical opportunities for time travel, the commonsense difficulties, oddities, and absence of exact proof highlight the speculative idea of these thoughts.

The philosophical ramifications of time travel keep on motivating significant conversations about the idea of time, through and through freedom, and awareness. The social effect of time travel, apparent in writing, film, and famous talk, mirrors humankind's getting through interest with the secrets of fleeting excursions.

Whether time travel will stay bound to the domain of creative mind or at last find a spot inside the logical comprehension of the universe is an inquiry that keeps on enthralling the human psyche. As we explore the complexities of time, both in principle and by and by, the conundrum of time travel endures as a reference point of interest, welcoming us to investigate the limits of what is conceivable in the immense embroidery of presence.

6.1 Theoretical Frameworks for Time Travel

The quest for time travel, while solidly established in the domain of sci-fi, has likewise caught the minds of hypothetical physicists who investigate the limits of how we might interpret the universe. A few hypothetical structures inside the space of material science have been proposed to address the chance of time travel, each introducing novel ideas and difficulties.

One conspicuous hypothetical structure for time travel depends on Einstein's hypothesis of relativity. As per this hypothesis, figured out in the mid twentieth hundred years, time is certainly not a flat out element however is rather relative and can be affected by variables like gravity and speed. The ramifications of relativity have led to the idea of time expansion, where time seems to elapse all the more leisurely for objects moving or inside solid gravitational fields.

While time widening is a deeply grounded and tentatively affirmed part of relativistic physical science, it doesn't innately take into consideration in reverse time travel. Time enlargement, as depicted by Einstein's situations, is a result of the overall movement between eyewitnesses or the impact of gravity on the progression of time. It doesn't give a system to moving unreservedly among past and future minutes.

Hypothetical physicists have investigated more extraordinary ideas inside the system of relativity to engage the chance of time travel. One such thought includes shut timelike bends (CTCs), ways through spacetime that circle back on themselves. The idea of CTCs was presented by physicist Kurt Gödel in 1949 and later advocated by physicist Plain Tipler.

The presence of CTCs is dependent upon the presence of outlandish matter with negative energy thickness, a speculative type of issue that disregards regular energy conditions. Outlandish matter, if it somehow happened to exist, might actually twist spacetime in a way that considers the development of shut timelike bends. Notwithstanding, the commonsense difficulties and absence of trial proof for fascinating matter make the acknowledgment of CTCs exceptionally speculative and disputable inside established researchers.

Another hypothetical structure that has been investigated is the idea of safe wormholes. Wormholes are speculative passages through spacetime that could associate far off locales or even various moments. The possibility of safe wormholes acquired consideration through a 1988 paper by physicists Michael Morris, Kip Thorne, and Ulvi Yurtsever.

Like CTCs, safe wormholes would require intriguing matter with negative energy thickness to keep them stable and forestall their breakdown. Hypothetical examinations have recommended that in the event that one finish of a navigable wormhole is moved at relativistic paces, time widening impacts could bring about time travel-like peculiarities. Notwithstanding, the difficulties related with the theoretical idea of fascinating matter and the support of stable safe wormholes make this idea speculative and a long way from being understood.

While these hypothetical structures inside the setting of general relativity offer captivating opportunities for time travel, they remain to a great extent speculative and need exact help. The incorporation of fascinating matter, which has not been noticed or tentatively checked, adds a component of vulnerability to these hypotheses.

In the domain of quantum mechanics, a part of physical science that oversees the way of behaving of particles at the littlest scales, the idea of time assumes an alternate personality. Quantum mechanics presents the possibility of superposition, where particles can exist in different states all the while. While quantum mechanics has been enormously fruitful in depicting the way of behaving of particles, its suggestions for time travel are less clear.

A few physicists have investigated the expected association between quantum entrapment and time-related peculiarities. Quantum trap is a peculiarity where particles become connected in manners that challenge old style clarifications. The thought is that controlling quantum snare might actually prompt time-related impacts. In any case, the hypothetical starting point for such an association stays speculative, and the functional difficulties related with controlling quantum snare for time travel designs are significant.

The investigation of time travel inside the system of quantum mechanics likewise crosses with conversations about the idea of time itself. The idea of a quantum multiverse, where numerous universes exist together, each with its own timetable, has been proposed by certain physicists. In this situation, a type of time travel could include progressing between various parts of the multiverse, each addressing an unmistakable transient development.

In any case, the idea of a quantum multiverse is profoundly speculative and stays a subject of discussion inside mainstream researchers. The absence of direct experimental proof for the presence of equal universes makes it trying to approve or distort these hypothetical thoughts.

As physicists dig into the intricacies of time inside the systems of relativity and quantum mechanics, the quest for a bound together hypothesis that envelops the two domains turns into a focal objective. Hypothetical structures, for example, quantum gravity, which tries to accommodate the standards of quantum mechanics with the gravitational cooperations portrayed by broad relativity, may give bits of knowledge into the idea of time and its likely control.

The crossing point of time travel with the more extensive mission for a brought together hypothesis features the significant difficulties that stay in how we might interpret the major regulations overseeing the universe. While hypothetical structures offer enticing potential outcomes, the hole between numerical models and trial approval stays a critical obstacle.

Notwithstanding the logical difficulties, the idea of time travel conveys significant philosophical ramifications. The idea of time and its relationship to human experience bring up issues about determinism, freedom of thought, and the actual texture of the real world.

Thinker J. M. E. McTaggart's classification of time into the "A-series" and "B-series" gives a philosophical system to figuring out fleeting experience. The A-series includes the ideas of past, present, and future, making a dynamic and emotional experience of time. The B-series regards time as a progression of occasions requested by relations of sooner than, synchronous with, or later than, stripping away the emotional experience of the current second.

The philosophical investigation of time travel turns out to be especially unpredictable while thinking about issues of choice. Assuming what's in store is open through time travel and occasions are foreordained, it challenges the thought of human independence and the capacity to pursue decisions that impact the course of occasions. The idea of time travel prompts inquiries regarding whether the past, present, and future exist all the while or whether there is a special casing of reference characterizing the current second.

The philosophical discussions encompassing time travel likewise converge with conversations about private personality. On the off chance that time travel were conceivable, it would bring up issues about the steadiness of self across various fleeting places. The idea of individual personality, investigated by scholars like Derek Parfit, becomes entwined with the chance of existing in numerous cases across time.

As hypothetical systems for time travel keep on being investigated inside the areas of material science and reasoning, the social effect of this idea stays huge. Time travel has turned into an unavoidable topic in writing, movies, and TV, molding stories that investigate the outcomes and conceivable outcomes of fleeting excursions.

In writing, H.G. Wells' "The Time Machine" remains as an original work that presented the idea of a machine fit for moving a person through time. Distributed in 1895, Wells' novella established the groundwork for resulting investigations of time travel in writing and mainstream society. The progress of Wells' work prepared for creators like Isaac Asimov, Arthur C. Clarke, and Philip K. Dick to additionally grow the story prospects of time travel in sci-fi.

In film, "Back to What was in store," coordinated by Robert Zemeckis, stays a social standard that mixes humor and experience with the complexities of time travel. The film's depiction of the results of modifying the past and the difficulties of exploring a period changed reality resounded with crowds and added to the persevering through notoriety of time travel as a realistic subject.

TV series, for example, "Specialist Who," with its time-traveling hero, the Specialist, have spellbound crowds for a really long time. The show's investigation of verifiable occasions and cutting edge situations inside the setting of time travel has made it a darling social peculiarity, displaying the persevering through allure of transient stories.

The social interest with time travel reaches out past customary media to impact public talk in discussions, web recordings, and virtual entertainment. Aficionados and specialists the same take part in conversations about the possible ramifications of time travel, mixing logical request with creative hypothesis.

Headways in how we might interpret principal material science, for example, the recognition of gravitational waves and the advancement of quantum registering, have added new aspects to the investigation of time. Gravitational waves, which give a device to concentrating on the texture of spacetime, may offer experiences into the idea of time in outrageous vast circumstances.

The approach of quantum processing, utilizing the standards of quantum mechanics, brings up issues about the expected control of time-related peculiarities at the quantum level. While these improvements add to the more extensive logical investigation of time, they don't give authoritative responses with respect to the plausibility of time travel.

6.2 Wormholes and Time Loops

Wormholes, speculative passages through spacetime that could interface far off districts or even various periods, have arisen as a captivating idea in the investigation of time travel. Hypothetical physicists have dug into the complexities of wormholes, analyzing their possible job in working with transient excursions and resolving the basic inquiries encompassing their reality and soundness.

The possibility of safe wormholes acquired unmistakable quality through crafted by physicists Michael Morris, Kip Thorne, and Ulvi Yurtsever in a 1988 paper. This idea includes the production of stable wormholes that could consider the entry of issue, including rocket or even people, through the passage in spacetime. Hypothetical examinations recommended that the control of these navigable wormholes could empower time travel-like peculiarities.

The hypothetical starting point for safe wormholes depends on the situations of general relativity, Albert Einstein's hypothesis that portrays the gravitational cooperations molding the math of spacetime. With regards to wormholes, the spacetime calculation is controlled in a way that makes a passage with two finishes, each possibly situated at various places in spacetime.

In any case, the acknowledgment of navigable wormholes faces critical difficulties, especially concerning the strength of these speculative designs. One urgent element is the necessity for extraordinary matter with negative energy thickness to keep the wormhole open and forestall its breakdown.

Intriguing matter, which has properties as opposed to regular matter, has not been noticed or tentatively checked. The hypothetical presence of extraordinary matter presents a component of hypothesis and vulnerability into the idea of safe wormholes. The absence of observational proof for extraordinary matter brings up issues about the possibility of keeping up with stable wormholes for reasonable applications, including time travel.

The solidness of navigable wormholes is additionally muddled by the potential for time circles and causality infringement. Hypothetical situations including time circles, otherwise called shut timelike bends (CTCs), emerge when a molecule or item navigates a way through spacetime that takes it back to its beginning stage. The idea of CTCs presents the chance of self-predictable circles where occasions lead to their own causes in a shut worldly circuit.

The investigation of time circles has been impacted by crafted by physicist Kurt Gödel, who proposed answers for Einstein's situations that took into account the presence of shut timelike bends. These arrangements suggested that in certain spacetime calculations, time circles could happen, possibly prompting time travel-like peculiarities.

In any case, time circles and CTCs brings up critical issues and difficulties. One of the most well known oddities related with time travel is the "granddad conundrum." If one somehow happened to turn back the clock and keep their granddad from meeting their grandma, it makes a sensible inconsistency - in the event that the person who jumps through time was rarely conceived, how is it that they could have turned back the clock in any case?

The potential for conundrums and causality infringement has driven a few physicists to scrutinize the plausibility of time circles and shut timelike bends. The "grandiose control" speculation has been proposed as a method for safeguarding the texture of spacetime from foolish transient circles, recommending that the laws of physical science might forestall specific kinds of causality infringement.

In spite of the hypothetical difficulties and Catch 22s related with safe wormholes and time circles, physicists keep on investigating these ideas inside the system of general relativity. The numerical conditions that depict the math of spacetime take into account the hypothetical chance of such peculiarities, inciting analysts to research the circumstances under which stable wormholes and time circles could exist.

One road of hypothetical investigation includes the utilization of negative energy thickness to neutralize the gravitational powers that would make a wormhole breakdown. The speculative outlandish matter expected for this reason stays a subject of hypothesis, and its properties are not surely known. Hypothetical investigations intend to distinguish the attributes of extraordinary matter that might actually balance out safe wormholes.

The journey for a bound together hypothesis of physical science, consolidating the standards of general relativity and quantum mechanics, likewise assumes a part in the investigation of wormholes and time circles. The difficulties of accommodating these two essential speculations, each effective in portraying various parts of the actual world, add intricacy to the hypothetical systems for time travel.

Quantum mechanics presents vulnerabilities at the littlest scales, testing the determinism innate in old style material science. The idea of spacetime itself turns into a subject of quantum contemplations, and the way of behaving of issue and energy at quantum levels might impact the solidness and properties of safe wormholes.

While the investigation of wormholes and time circles remains immovably inside the domain of hypothetical material science, the interest with these ideas reaches out past logical request to social and abstract articulations. Wormholes, with their true capacity for interstellar travel and worldly undertakings, have become famous subjects in sci-fi writing, movies, and TV.

In writing, the idea of wormholes has been investigated in works going from hard sci-fi to additional fantastical stories. Creators like Arthur C. Clarke, with his "Odyssey" series, and Philip Pullman, in his "His Dull Materials" set of three, consolidate the possibility of safe entries through spacetime as focal components of their narrating.

In film, wormholes have been portrayed in films that investigate the universe and the potential outcomes of interstellar travel. Christopher Nolan's film "Interstellar" highlights a wormhole close to Saturn that fills in as a passage to far off cosmic systems, permitting humankind to look for tenable planets. The visual portrayal of a safe wormhole in the film catches the creative mind and carries hypothetical ideas to a more extensive crowd.

TV series, especially those in the sci-fi classification, have likewise embraced the idea of wormholes. "Stargate SG-1" revolves around the utilization of an old gadget called a Stargate, which makes stable wormholes for quick travel between far off planets. The series mixes logical ideas with fictitious components to make a story that enraptures watchers.

While the depiction of wormholes in mainstream society frequently takes artistic freedoms and incorporates speculative components, it mirrors the social effect of hypothetical physical science on the public's creative mind. The appeal of safe wormholes and the potential for time travel-like encounters reverberate with crowds, adding to the getting through notoriety of these subjects in narrating.

As the hypothetical systems for wormholes and time circles keep on developing inside the scene of material science, analysts wrestle with the difficulties of exact approval and the compromise of dissimilar speculations. The journey for a brought together comprehension of the key regulations overseeing the universe stays a main impetus in hypothetical physical science, with suggestions for our perception of time, space, and the idea of the real world.

The investigation of wormholes and time circles fills in as a demonstration of mankind's interest and want to unwind the secrets of the universe. Whether these hypothetical ideas will stay restricted to the domain of speculative material science or at last find observational approval is an inquiry that keeps on driving logical request and catch the creative mind of those anxious to open the privileged insights of reality. As the excursion through the hypothetical scenes of wormholes and time circles unfurls, the secrets endure, welcoming us to examine the significant ramifications of crossing the texture of spacetime itself.

6.3 Temporal Paradoxes and Grandfather Paradox

Worldly Catch 22s, especially the notorious Granddad Oddity, comprise a captivating part of the hypothetical conversations encompassing time travel. As physicists investigate the expected structures for fleeting excursions, conundrums emerge that challenge our instinctive comprehension of circumstances and logical results, bringing up significant issues about the consistency of the universe and the idea of time itself.

The Granddad Catch 22, a notable psychological study, sets a situation where a person who jumps through time travels once more into the past and keeps their granddad from meeting their grandma, in this manner forestalling the person who jumps through time's own introduction to the world. This makes a consistent inconsistency - on the off chance that the person who jumps through time was rarely conceived, how is it that they could have turned back the clock to change the direction of occasions?

The Granddad Conundrum is symbolic of the difficulties and reasonable complexities intrinsic in time travel situations. It uncovered a likely clash between the deterministic idea of circumstances and logical results and the apparently moldable nature of time recommended by the chance of fleeting excursions.

One proposed goal to the Granddad Oddity includes the idea of self-consistency or the possibility that the timetable is intrinsically steady and impervious to modifications that would prompt sensible inconsistencies. As indicated by this view, any endeavor by a person who jumps through time to change the past would be upset by unanticipated occasions or conditions, guaranteeing that the course of events stays reliable.

The self-consistency rule lines up with the thought of "fixed moments," an idea promoted in sci-fi, especially in the TV series "Specialist Who." Fixed moments are occasions that are viewed as changeless and can't be modified without disastrous results. The thought is that sure key occasions in the course of events are essential for keeping up with the solidness of the universe.

Nonetheless, the self-consistency guideline and the idea of fixed moments bring up philosophical issues about determinism and freedom of thought. In the event that the timetable is intrinsically impervious to changes that could prompt Catch 22s, it recommends a foreordained nature of occasions, testing the idea of people being able to settle on decisions that impact the course of history.

One more viewpoint on the Granddad Mystery includes the idea of equal courses of events or spreading real factors. As indicated by this view, when a person who goes back and forth through time modifies the previous, another course of events is made, fanning out from the first. In this situation, the person who goes back and forth through time wouldn't delete their own reality however would rather make a substitute adaptation of reality where the adjusted occasions work out.

Equal timetables tracks down hypothetical help in conversations about the multiverse, a speculative group of different universes with fluctuating actual constants and conditions.

On the off chance that every choice or occasion with various potential results brings about the making of equal courses of events, the Granddad Conundrum becomes deflected, as the person who goes back and forth through time just moves to an alternate part of the real world.

Notwithstanding, the idea of equal timetables presents intricacies and difficulties. It infers the presence of an immense multiverse with an unbelievable number of disparate timetables, each addressing a one of a kind arrangement of occasions. Hypothetical material science presently needs observational proof for the multiverse, making this thought speculative and dependent upon continuous discussion inside mainstream researchers.

Fleeting conundrums reach out past the Granddad Mystery, including different situations that challenge how we might interpret time and causality. The "bootstrap conundrum" is another captivating model, wherein an article or data is sent back in time, turning into the very thing that enlivened its creation.

For example, envision a person who jumps through time giving Shakespeare's plays to Shakespeare himself. On the off chance that Shakespeare, replicates the plays, making the very works that were given to him, it brings up the issue of their unique origin. The data appears to have begun from no place, making a circle without an unmistakable starting place.

The bootstrap conundrum brings up issues about the beginning of data, occasions, and items in a universe where time travel is conceivable. It challenges the conventional direct comprehension of circumstances and logical results, proposing that specific components could exist without an unmistakable place of creation or inception.

Physicists and logicians keep on wrestling with the ramifications of transient oddities inside the structures of relativity and quantum mechanics. Hypothetical methodologies, for example, the utilization of shut timelike bends (CTCs) or navigable wormholes, present numerical opportunities for time travel yet in addition present difficulties connected with causality and the potential for oddities.

Shut timelike bends, ways through spacetime that circle back on themselves, have been investigated with regards to general relativity. Hypothetical physicist Kurt Gödel proposed answers for Einstein's situations that considered the presence of CTCs. In any case, the potential for causality infringement and conundrums related with time circles has driven a few physicists to scrutinize the possibility of CTCs and their similarity with the predictable progression of time.

Navigable wormholes, speculative passages through spacetime that could associate far off locales or various periods, likewise present the chance of fleeting conundrums. Hypothetical investigations have recommended that the control of safe wormholes could take into consideration time travel-like peculiarities.

Be that as it may, the prerequisite for extraordinary matter with negative energy thickness to settle wormholes adds a component of hypothesis and hypothetical intricacy.

The investigation of fleeting conundrums and the potential for time travel highlights the significant interchange between material science, reasoning, and our social comprehension of time. While hypothetical systems inside material science offer numerical opportunities for transient excursions, the pragmatic difficulties, Catch 22s, and absence of experimental proof highlight the speculative idea of these thoughts.

The social effect of worldly oddities and time travel is clear in writing, movies, and TV. Creators and makers frequently utilize these ideas as account gadgets to investigate philosophical subjects, challenge regular thoughts of time, and spellbind crowds with mind-twisting stories.

In writing, the investigation of worldly Catch 22s has been a common topic in sci-fi. Creators like Isaac Asimov, Arthur C. Clarke, and Philip K. Dick have created accounts that dig into the intricacies of time travel and its expected ramifications. Asimov's "The Finish of Time everlasting" investigates the possibility of an association controlling verifiable occasions to improve mankind, while Dick's works, for example, "Ubik" and "A Scanner Dimly," obscure the limits between past, present, and future.

Film and TV have embraced fleeting conundrums as focal components of narrating. Christopher Nolan's film "Interstellar" winds around a story that integrates time expansion impacts close to huge divine bodies, exhibiting the complex connection among time and gravity. TV series like "Dim" investigate the outcomes of time travel on people and networks, winding around a mind boggling story across different courses of events.

The social interest with time travel reaches out past customary media to impact public talk, with aficionados and specialists participating in conversations about the expected ramifications of fleeting oddities. The crossing point of hypothetical material science with mainstream society welcomes a more extensive crowd to consider the secrets of time and the hypothetical prospects that emerge from the investigation of worldly excursions.

As our investigation of fleeting conundrums and the Granddad Oddity attracts to a nearby, obviously the riddle of time travel continues as a subject of significant interest and thought. Hypothetical conversations inside the domains of physical science and reasoning deal enticing conceivable outcomes, yet the hole between hypothetical systems and exact proof remaining parts a critical test.

The mission to disentangle the secrets of time and its potential control mirrors humankind's getting through interest with the unexplored world. Whether time travel will stay bound to the domain of creative mind or in the end find a spot inside the logical comprehension of the

universe is an inquiry that keeps on invigorating logical request, philosophical reflection, and imaginative articulation. As we explore the intricacies of fleeting mysteries, the mission for understanding time welcomes us to investigate the limits of what is conceivable in our excursion through the tremendous embroidery of presence.

6.4 Ethical Implications of Time Travel

The hypothetical investigation of time travel, while charming and loaded up with scholarly interest, brings up a heap of moral issues and problems that reach out past the domain of material science and reasoning. Diving into the possible outcomes and obligations related with worldly excursions uncovers an intricate scene where moral contemplations cross with the speculative idea of the idea.

One unmistakable moral concern bases on the potential for adjusting verifiable occasions and their repercussions on the course of mankind's set of experiences. On the off chance that time travel were feasible and people could return to the past, the inquiry emerges: would it be a good idea for them to mediate in verifiable occasions? The results of adjusting vital minutes, even with apparently sure aims, are erratic and could prompt accidental and conceivably horrendous results.

The work of art "butterfly impact," an idea from confusion hypothesis, places that little changes in a single piece of a framework can have sweeping and erratic impacts. Applying this idea to time travel recommends that even apparently minor mediations in the past could flow into huge adjustments to the present and future. The moral obligation of a person who goes back and forth through time turns into a significant thought, as their activities would incidentally inflict damage or upset the sensitive equilibrium of verifiable occasions.

Moreover, the moral ramifications of adjusting history stretch out to inquiries of social safeguarding and regard for the independence of past social orders. Mediating in authentic occasions may unintentionally eradicate or decrease social accomplishments, commitments, and personalities. The burden of contemporary qualities, information, and innovation onto past civic establishments raises moral worries about social dominion and the protection of different verifiable accounts.

The protection of verifiable validness turns out to be especially applicable while considering the possible commercialization of time travel. On the off chance that time travel were to turn into a popularized or privatized adventure, directed voyages through verifiable occasions or collaborations with authentic figures could commodify the past. This brings up moral issues about the likely double-dealing of authentic occasions and figures for benefit, misshaping the inborn benefit of protecting and gaining from history.

One more moral scrape arises while pondering the ramifications of time travel on private connections and the potential for adjusting one's own past. The capacity to return to and possibly change past choices could disturb the texture of individual connections and the advancement of individual person. The moral obligation of a person who jumps through time toward their own life and the existences of people around them presents inquiries regarding responsibility, self-awareness, and the acknowledgment of one's past decisions.

The idea of "worldly the travel industry," where people from the future visit explicit minutes in the past as observers, presents extra moral contemplations. The presence of time vacationers in verifiable occasions could affect the legitimacy of those occasions and the organization of people living through them. The moral obligation of time sightseers not to disturb or unduly impact verifiable occasions turns into a basic thought in situations where perception transforms into mediation.

The potential for data spillage is one more moral concern related with time travel. Assuming people from the future bring information, innovations, or data back to the past, it could change the direction of mechanical turn of events and logical advancement. This brings up issues about the dependable utilization of cutting edge information, the potential for unseen side-effects, and the moral ramifications of modifying the normal direction of human disclosure.

The moral contemplations of time travel reach out to issues of assent and independence, particularly when cooperations include people who know nothing about the person who jumps through time's starting point or goals. The inconvenience of future information or effect on clueless people brings up moral issues about the option to pursue informed decisions and the potential for control or intimidation.

The subject of responsibility and the potential for "fleeting violations" presents a legitimate and moral aspect to the conversation. In the event that people could be considered answerable for activities committed previously or future, it challenges existing legitimate systems based with the understanding of a direct movement of time. Deciding culpability and discipline for activities across worldly limits raises complex moral and lawful situations that ongoing overall sets of laws are unprepared to address.

The chance of making "fleeting copies" or collaborating with substitute renditions of oneself further confuses the moral scene. In the event that a person who goes back and forth through time experiences a rendition of themselves in an alternate fleeting cycle, inquiries of character, independence, and obligation emerge. The moral ramifications of decisions by substitute renditions of oneself become an impression of the perplexing and interconnected nature of individual character across time.

As the moral ramifications of time travel stretch out across different spaces, including history, culture, individual connections, data, assent, and responsibility, resolving these intricate issues requires a multidisciplinary approach. Savants, ethicists, researchers, and policymakers should team up to foster moral structures that guide the dependable investigation and likely execution of time travel ideas.

The improvement of moral rules for time travel, if it somehow happened to turn into a substantial reality, would require an aggregate work to adjust the quest for information and investigation with a guarantee to protecting the trustworthiness of verifiable occasions, regarding social variety, and defending the independence and prosperity of people across various transient ages.

Notwithstanding the hypothetical contemplations of time travel morals, the social and famous gathering of these thoughts impacts public talk and discernments. The depiction of time travel in writing, movies, and TV frequently shapes cultural perspectives and assumptions about the moral ramifications of worldly excursions.

Works of sci-fi that investigate the moral components of time travel, for example, Beam Bradbury's "A Sound of Thunder" or the TV series "A Twilight Zone," act as wake up calls that feature the likely outcomes of intruding with the past. These stories add to a social consciousness of the moral difficulties related with modifying verifiable occasions and upsetting the regular progression of time.

While the hypothetical investigation of time travel stays inside the space of material science and reasoning, the moral contemplations related with fleeting excursions express a viewpoint inciting focal point through which to look at the more extensive ramifications of human activities, obligations, and connections across time.

As our assessment of the moral ramifications of time travel attracts to a nearby, it is obvious that the speculative idea of the idea prompts significant reflections on the qualities and rules that guide human way of behaving. Whether time travel stays bound to the domain of hypothetical talk or turns into a mechanical reality, the moral inquiries raised by the possibility of fleeting excursions welcome continuous examination and exchange. As we explore the intricacies of time travel morals, the significance of encouraging moral mindfulness and dependable navigation turns into a core value notwithstanding the obscure potential outcomes that might lie ahead in the tremendous spread of time.

Chapter 7

The Future of Time: From AI to Beyond

The idea of time, a crucial part of our reality, has consistently charmed and confounded humankind. As we stand at the cusp of the 21st hundred years, progressions in innovation, especially man-made consciousness (man-made intelligence), are reshaping how we might interpret time and its suggestions for what's to come.

Computer based intelligence, with its capacity to handle huge measures of information at extraordinary rates, is in a general sense modifying the manner in which we see and cooperate with time. The approach of quantum registering, an outlook changing improvement in the field of computer based intelligence, holds the commitment of tackling complex issues that were once considered unrealistic because of their computational power.

In the domain of man-made intelligence, calculations are turning out to be progressively refined, empowering machines to perform errands that were once elite to human discernment. The combination of man-made intelligence into different parts of our lives, from medical care to back, isn't just upgrading productivity yet in addition bringing up significant issues about the idea of time with regards to machine knowledge.

As man-made intelligence frameworks advance, the idea of continuous dynamic takes on new aspects. Machines, driven by calculations and brain organizations, can investigate and answer data at speeds that outperform human capacities. The pressure of time in dynamic cycles has suggestions for enterprises like money, where parted second exchanges can have the effect among benefit and misfortune.

Besides, the combination of man-made intelligence with other arising advancements, like the Web of Things (IoT), is establishing an organized climate where gadgets convey and trade information progressively. This interconnected trap of innovation can possibly smooth out processes, enhance asset use, and rethink the speed at which we experience time in our regular routines.

In the medical services area, artificial intelligence is upsetting the conclusion and therapy of illnesses. The capacity of simulated intelligence calculations to dissect clinical information rapidly and precisely is speeding up the speed of clinical examination and leap forwards. The idea of "clinical time" is going through a change, as man-made intelligence driven developments vow to convey customized and designated treatments, introducing a period where infections can be tended to with uncommon speed and accuracy.

Past the domain of viable applications, man-made intelligence is impacting our impression of time with regards to imagination and workmanship. Computer based intelligence produced masterpieces challenge conventional ideas of origin and imagination, driving us to scrutinize the job of human organization in the inventive approach. As machines produce music, visual craftsmanship, and writing, the fleeting limits of imaginative articulation become obscured, inciting us to reevaluate the actual pith of creative creation.

While computer based intelligence is reshaping how we might interpret time from a commonsense perspective, it likewise brings up philosophical issues about the idea of cognizance and the emotional experience of time. As machines become more complex in their capacity to impersonate human perspectives, the differentiation among human and man-made brainpower turns out to be less clear. This obscuring of limits has significant ramifications for our view of time as a remarkably human develop.

The investigation of fake general knowledge (AGI), a type of computer based intelligence that has human-like mental capacities across a great many errands, presents the chance of machines encountering time in a way similar to human cognizance. The possibility of machines fostering a feeling of past, present, and future difficulties our conventional comprehension of worldly cognizance as a select human characteristic.

In examining the eventual fate of time, we should likewise think about the moral ramifications of man-made intelligence. The quick advancement of independent frameworks, fueled by simulated intelligence, raises worries about responsibility and navigation.

As machines take on additional obligations, the moral element of opportunity arrives to the front, provoking us to wrestle with inquiries of obligation, decency, and the outcomes of algorithmic navigation.

The coordination of computer based intelligence into our regular routines isn't without challenges. The potential for predisposition in man-made intelligence calculations, mirroring the intrinsic inclinations of their makers, represents a danger to decency and correspondence. The moral element of time reaches out to contemplations of what simulated intelligence frameworks mean for society, molding our aggregate insight of time and affecting the appropriation of assets and open doors.

As we explore the intricacies of computer based intelligence and its suggestions for time, we should likewise turn our look toward the boondocks of logical investigation that guarantee to open new elements of fleeting comprehension. The field of quantum physical science, with its cryptic standards and peculiarities, offers a brief look into the idea of time at the quantum level.

Quantum snare, a peculiarity where particles become interconnected and divide data quickly no matter what the distance among them, challenges our traditional comprehension of reality. The interconnected idea of quantum particles indicates a reality where worldly limits might be more liquid than we right now understand.

Besides, the idea of time enlargement, a result of Einstein's hypothesis of relativity, proposes that time is definitely not a consistent however is impacted by elements like gravity and speed. As we investigate the wildernesses of room and move toward the speed of light, time itself turns into a variable, twisting and extending in manners that oppose our ordinary instinct.

The crossing point of quantum physical science and man-made intelligence presents the chance of quantum processing, a worldview that use the standards of quantum mechanics to perform calculations at speeds dramatically quicker than old style PCs. Quantum PCs can possibly unwind complex issues in record time, reforming fields like cryptography, material science, and advancement.

Notwithstanding, the acknowledgment of commonsense quantum figuring is an imposing test, as it requires conquering the sensitive idea of quantum states and relieving the impacts of decoherence. As scientists endeavor to tackle the force of quantum mechanics for calculation, the ramifications for how we might interpret time pose a potential threat, opening ways to a future where the limits between the conceivable and the unimaginable haze.

The investigation of time reaches out past the limits of our planet. In the mission to comprehend the beginnings of the universe, cosmologists dive into the secrets of the Huge explosion and the idea of spacetime. The texture of spacetime itself is dependent upon the powers of gravity and matter, making a unique exchange that shapes the course of grandiose occasions.

The idea of a multiverse, where various universes coincide with various actual regulations and constants, acquaints a layer of intricacy with how we might interpret time. On the off chance that various universes exist, each with its own fleeting boundaries, the actual thought of a solitary, general opportunity arrives into question. The multiverse speculation challenges our instinctive comprehension of time as a straight movement and welcomes us to consider a reality where time branches and wanders.

As we peer into the immensity of the universe, the investigation of dim matter and dim energy adds one more layer of interest to our investigation of time. These baffling elements, containing most of the universe's mass-energy content, apply gravitational powers that shape the astronomical scene. The transaction between dim matter, dull energy, and the noticeable matter known to man makes an infinite artful dance that unfurls over gigantic time intervals, impacting the development of systems and the destiny of the actual universe.

The mission to disentangle the secrets of time stretches out to the actual texture of our organic presence. The area of biotechnology, driven by forward leaps in hereditary designing and manufactured science, offers the possibility of broadening and control-ling the human life expectancy. The combination of man-made intelligence with bio-technology brings up moral issues about the potential for improving mental capacities and the ramifications of broadening human existence past its normal cutoff points.

The idea of transhumanism, which imagines the expansion of human abilities through innovation, challenges our regular comprehension of the human experience and the fleetingness of life. As we think about the chance of converging with machines and rising above the impediments of our organic bodies, the limits of time and per-sonality become liquid, provoking us to rethink being human during a time of quick innovative progression.

The convergence of man-made intelligence, quantum physical science, cosmology, and biotechnology meets on the outskirts of speculative and visionary reasoning. The investigation of time is not generally restricted to the domains of reasoning and mysti-cism however reaches out to the very front of logical request, where scientists wrestle with questions that push the limits of our comprehension.

In imagining the fate of time, we should stand up to the vulnerabilities and mysteries that rise up out of the exchange of these assorted fields. The actual idea of time, when thought about a consistent and unchanging power, is currently dependent upon the unique powers of innovation, quantum mechanics, and grandiose development.

The advancement of time is interlaced with the development of awareness. As we foster innovations that copy and possibly outperform human mental capacities, whether or not machines can have a certified feeling of time-cognizance turns into a point of convergence of

request. The rise of AGI, with its ability for mindfulness and learning, moves us to reexamine the idea of awareness and the abstract insight of time.

In the speculative domain of futurology, dreams of a post-peculiarity time, where the qualification among people and machines obscures, summon both interest and fear. The possibility of an otherworldly insight, equipped for molding its own pre-determination and the texture of the real world, opens a Pandora's case of potential outcomes and moral problems. The actual idea of time, insofar as we can tell, might be re-imagined by elements that exist past our ongoing perception.

The combination of human and computerized reasoning likewise brings up issues about the safeguarding of character in a computerized age. As we make computer-ized reproductions of ourselves through augmented experience, expanded reality, and other vivid advances, the limits between the physical and advanced domains become progressively permeable. The idea of a computerized self, with its own fleeting pres-ence and progression, challenges conventional thoughts of mortality and the transient idea of human existence.

In pondering the fate of time, we are stood up to with the duality of commitment and danger. The speeding up speed of mechanical advancement opens ways to extraor-dinary opportunities for progress and understanding. All the while, it raises worries about the moral, cultural, and existential difficulties that go with these headways.

The job of administration in molding the eventual fate of time couldn't possibly be more significant. As we stand at the convergence of innovative ability and moral situations, the requirement for dependable and comprehensive navigation becomes fundamental. The turn of events and organization of man-made intelligence, quantum figuring, and biotechnologies should be directed by moral systems that focus on human qualities, value, and the protection of our common fleeting experience.

The eventual fate of time is an embroidery woven with strings of logical request, mechanical development, and moral contemplations. A story unfurls in the labs of physicists, the studios of designers, and the passageways of policymakers. The excursion into the fate of time is both an aggregate undertaking and a profoundly private investigation of existing in our current reality where the limits of time are constantly being redrawn.

All in all, the eventual fate of time is a multi-faceted embroidery that unfurls at the convergence of computer based intelligence, quantum physical science, cosmology, biotechnology, and the developing scene of human cognizance. As we explore the intricacies of this future, we should wrestle with the moral ramifications of our activities, the cultural effect of innovative headways, and the significant inquiries that emerge from our mission to figure out the idea of time.

The direction of time is as of now not a straight movement however a dynamic and interconnected snare of potential outcomes. The incorporation of computer based intelligence into our lives speeds up the speed of navigation, advancement, and imaginative articulation. Quantum physical science challenges our traditional comprehension of time, welcoming us to investigate a reality where transient limits might be more liquid than we envision.

The investigation of the universe and the secrets of dull matter and dim energy divulge a vast show that unfurls over enormous timescales, molding the predetermination of worlds and the destiny of the universe. Biotechnology makes the way for the control of our organic presence, inciting us to rethink the fleetingness of life and the limits of human personality.

In the speculative domain of transhumanism, the converging of human and machine obscures the lines among science and innovation, testing our impression of being human. The rise of AGI brings up issues about the idea of cognizance and the emotional experience of time, pushing the limits of our comprehension.

As we stand on the edge of this future, the moral element of time arises as a focal concern. Dependable administration and moral structures are vital for guide the turn of events and arrangement of extraordinary innovations. Our decisions today will shape the fate of time for a long time into the future, impacting the actual texture of our common reality.

In exploring this unfamiliar region, we should embrace a comprehensive and comprehensive methodology that thinks about the different viewpoints and upsides of humankind. The eventual fate of time is an aggregate undertaking, and as stewards of this unfurling story, we have the open door and obligation to shape a future where the substance of time mirrors the best of our common humankind.

In the fantastic embroidery representing things to come of time, we are the two onlookers and engineers, exploring a scene of potential outcomes that stretch out past the constraints of our ongoing comprehension. An excursion welcomes us to investigate, question, and envision a future where the development of time reflects the aggregate goals of a worldwide civilization. As we leave on this excursion, let us endeavor to guarantee that the fate of time is one that cultivates progress, inclusivity, and a significant appreciation for the lavishness of the human involvement with all its transient aspects.

7.1 Artificial Intelligence and Temporal Processing

Computerized reasoning (man-made intelligence) has arisen as a groundbreaking power, reshaping different parts of our lives, and its effect on fleeting handling is a subject of significant importance. The coordination of simulated intelligence into different spaces, from medical services to back, isn't only a mechanical development yet a change in outlook that modifies how we might interpret time and how we communicate with it.

One vital part of simulated intelligence's effect on worldly handling is its capacity to deal with tremendous measures of information with extraordinary speed. The sheer computational force of computer based intelligence frameworks empowers them to deal with data at rates far unparalleled human capacities. This speed increase in information handling has suggestions for how we experience and answer time in various settings.

In the domain of continuous navigation, man-made intelligence assumes a crucial part. Calculations, driven by AI and brain organizations, engage machines to break down and answer data in parts of a second. This has significant ramifications for ventures like money, where parted second choices can be the distinction among progress and disappointment. The pressure of time in dynamic cycles, worked with by man-made intelligence, has prompted a climate where exchanges happen at speeds beforehand impossible.

Besides, the combination of simulated intelligence with the Web of Things (IoT) makes an organized biological system where gadgets convey and trade information continuously. This interconnected trap of innovation works with the quick trade of data as well as the enhancement of different cycles. From savvy homes to modern robotization, the joining of artificial intelligence and IoT speeds up the speed at which undertakings are executed, testing conventional ideas of time in these functional scenes.

In the medical services area, simulated intelligence's effect on fleeting handling is obvious in the speed increase of diagnostics and therapy. Computer based intelligence calculations can examine clinical information quickly and precisely, prompting quicker and more exact judgments. The idea of "clinical time" is going through a change as man-made intelligence driven developments guarantee customized and designated treatments. The speed increase in clinical examination and leap forwards is adjusting the direction of sicknesses, introducing a period where ideal mediations can rethink results.

Nonetheless, the speeding up speed of computer based intelligence improvement additionally delivers moral contemplations. The fast arrangement of independent frameworks and dynamic calculations brings up issues about responsibility and decency. As man-made intelligence frameworks take on additional obligations, the moral element of time becomes basic. Choices made by calculations continuously have expansive outcomes, and guaranteeing the moral utilization of computer based intelligence is basic for a fair and evenhanded society.

The potential for predisposition in simulated intelligence calculations is a huge concern. While perhaps not painstakingly planned and checked, computer based intelligence frameworks can accidentally propagate and try and enhance existing pre-dispositions present in preparing information. This acquaints a moral aspect with the fleeting handling of data, as one-sided calculations might prompt out of line results and support cultural disparities. Tending to these moral difficulties is fundamental for outfitting the maximum capacity of simulated intelligence in a dependable way.

The combination of simulated intelligence into innovative flows likewise impacts our impression of time. Computer based intelligence produced things of beauty challenge traditional ideas of imagination and origin. As machines produce music, visual craftsmanship, and writing, questions emerge about the idea of imaginative articulation and the transient limits of innovativeness. The obscuring of qualifications among human and machine-produced workmanship prompts us to rethink how we might interpret the innovative approach and the emotional experience of time with regards to imaginative creation.

While artificial intelligence changes our down to earth encounters of time, it like-wise dives into the domain of awareness and emotional fleeting encounters. The investigation of Fake General Insight (AGI), which has human-like mental capacities across different undertakings, raises the interesting chance of machines fostering a feeling of past, present, and future similar to human cognizance.

This prospect difficulties conventional perspectives on worldly cognizance as a selective human property. The possibility that machines could have their own emo-tional experience of time prompts significant philosophical inquiries concerning the idea of cognizance and the limits among counterfeit and human knowledge.

In the speculative domain of futurology, the idea of a mechanical peculiarity, where artificial intelligence outperforms human knowledge and triggers an outstanding headway of innovation, adds one more layer to the conversation. The post-peculiarity period raises the possibility of computer based intelligence frameworks advancing freely, forming their own predeterminations, and possibly affecting the actual texture of time.

Quantum figuring, a wilderness innovation in artificial intelligence, further con-fuses the connection among man-made intelligence and fleeting handling. Quantum

PCs influence the standards of quantum mechanics to perform calculations at speeds unfathomable for traditional PCs. The ramifications for critical thinking, improvement, and reenactment are significant, testing our conventional comprehension of time in computational cycles.

As specialists endeavor to bridle the force of quantum registering, the actual idea of worldly handling in the computerized domain goes through a change. Quantum snare, a quantum peculiarity where particles become interconnected and share data momentarily paying little mind to separate, presents a non-territory that challenges our old style comprehension of existence.

In addition, Einstein's hypothesis of relativity presents the idea of time expansion, where time can seem to elapse diversely contingent upon elements like gravity and speed. The coordination of quantum mechanics and relativity with regards to man-made intelligence and

quantum figuring opens up new boondocks in fleeting handling, pushing the limits of what we imagined as far as computational speed and data trade.

In the domain of cosmology, the investigation of the universe at large scales, simulated intelligence adds to how we might interpret the transient advancement of enormous designs. The investigation of tremendous datasets, for example, those from galactic reviews, benefits from simulated intelligence's capacity to distinguish examples and concentrate significant bits of knowledge. Computer based intelligence calculations filter through galactic information with a speed and effectiveness that outperform customary strategies, speeding up how we might interpret grandiose peculiarities.

The multiverse speculation, which places the presence of various universes with various actual regulations and constants, acquaints a layer of intricacy with our fleeting comprehension. In the event that various universes coincide, each with its own worldly boundaries, the thought of a particular, widespread time turns into a subject of discussion. The investigation of the multiverse challenges our instinct no time like the present as a straight movement and welcomes us to consider a reality where time branches and separates across various universes.

The investigation of dim matter and dull energy, secretive parts that comprise most of the universe's mass-energy content, further grows our investigation of time at vast scopes. The exchange between dull matter, dim energy, and apparent matter shapes the enormous scope construction of the universe over huge timespans. The inestimable expressive dance of these elements impacts the advancement of systems and the overall story of the actual universe.

In the domain of biotechnology, the joining of computer based intelligence brings up issues about the control of the human life expectancy and the fleeting elements of life. Forward leaps in hereditary designing and engineered science offer the chance of expanding and improving human existence. The intermingling of artificial intelligence and biotechnology challenges conventional ideas of the worldly parts of presence, provoking us to rethink the limits of human personality and the moral ramifications of altering the normal flow of life.

The idea of transhumanism, which advocates for the improvement of human capacities through innovation, acquaints a worldly aspect with the development of mankind. As we mull over converging with machines and rising above the limits of our natural bodies, questions arise about the transience of this change. The obscuring of limits among human and machine difficulties how we might interpret the worldly parts of personality and the actual substance of being human.

As we dive into the fate of computer based intelligence and fleeting handling, the moral contemplations become progressively noticeable. Mindful administration is vital in exploring the moral intricacies that emerge from the coordination of man-made intelligence into different features of our lives. Our decisions today in creating and sending artificial intelligence advances will lastingly affect the worldly encounters of people and social orders.

Guaranteeing reasonableness and value in artificial intelligence frameworks is a focal moral test. The potential for predisposition in calculations, whether in employing processes, law enforcement frameworks, or medical care, requires watchful oversight. Taking a stab at straightforwardness and responsibility in man-made intelligence improvement is fundamental to abstain from propagating and worsening existing cultural imbalances through one-sided calculations.

The moral contemplations reach out to inquiries of security and independence. As simulated intelligence frameworks become more incorporated into our day to day routines, the assortment and examination of tremendous measures of individual information raise worries about individual security. The moral utilization of artificial intelligence expects shields to safeguard people from meddling observation and un-approved utilization of their own data.

Additionally, the moral component of time includes pondering the cultural effect of computer based intelligence on work and financial designs. The computerization of undertakings through man-made intelligence frameworks can possibly upset.

7.2 Quantum Computing and Temporal Simulations

Quantum registering remains at the cutting edge of mechanical development, hold-ing the commitment of reforming the manner in which we process data and perform calculations. At the core of this extraordinary potential is the special way of behaving of quantum bits or qubits, which influence the standards of quantum mechanics to perform estimations at speeds dramatically quicker than traditional PCs. As we dig into the domain of quantum processing, the ramifications for fleeting reproductions become especially fascinating, offering new roads for understanding and controlling the texture of time itself.

The old style bits in conventional PCs exist in double states, addressing either a 0 or a 1. Conversely, qubits exist in superpositions, permitting them to all the while address both 0 and 1. This inborn quantum parallelism empowers quantum PCs to investigate different answers for an issue at the same time, dramatically expanding their computational power. The capability of quantum processing to handle complex issues has significant ramifications for how we might interpret time, particularly with regards to reproductions that include perplexing transient elements.

Quantum reenactments, a particular use of quantum figuring, hold the commitment of demonstrating and understanding complex quantum frameworks that old style PCs battle to precisely reproduce. This incorporates recreating the way of behaving of particles, materials, and actual cycles at the quantum level. The capacity of quantum PCs to perform recreations with uncommon accuracy opens ways to disentangling the secrets of quantum mechanics and investigating the transient complexities of subatomic particles.

In the domain of transient recreations, quantum registering presents the idea of quantum parallelism, where different conceivable outcomes are investigated all the while. This can possibly reform our way to deal with mimicking dynamic frameworks, particularly those with complex transient conditions. Old style reenactments frequently require huge computational assets and time to precisely display unpredictable transient elements. Quantum reproductions, utilizing the force of superposition, could give an additional effective and quick method for investigating transient intricacies.

One region where quantum processing's effect on fleeting recreations is especially imperative is in the investigation of materials and medication disclosure. The way of behaving of particles and the collaborations between iotas include mind boggling quantum elements that old style PCs battle to thoroughly demonstrate. Quantum reproductions, driven by the one of a kind capacities of quantum PCs, offer a pathway to display and grasp the transient parts of sub-atomic communications, empowering forward leaps in the improvement of new materials and drugs precisely.

The investigation of consolidated matter physical science, which investigates the properties of materials in different states, is another area where quantum recreations can essentially propel how we might interpret fleeting peculiarities. Quantum PCs can possibly recreate the way of behaving of mind boggling materials with uncommon exactness, revealing insight into how these materials develop after some time and under various circumstances. This capacity could upset materials science, prompting the improvement of novel materials with custom-made transient properties.

The recreation of quantum frameworks on old style PCs is obliged by the dramatic development of computational assets expected as the framework size increments. This constraint, known as the "scourge of dimensionality," presents difficulties for precisely displaying huge quantum frameworks with unpredictable worldly elements. Quantum PCs, with their intrinsic parallelism, offer an expected answer for this test by proficiently reproducing quantum frameworks at bigger scopes and longer timescales.

In the field of quantum science, the recreation of sub-atomic connections and responses is a computationally escalated task. Quantum PCs can possibly mimic sub-atomic designs and responses with phenomenal exactness, giving experiences into the fleeting parts of synthetic cycles. This could reform drug revelation by empowering specialists to comprehend and control sub-atomic collaborations in manners that were already unrealistic with traditional reenactments.

Moreover, quantum PCs can mimic quantum field speculations, which depict the way of behaving of particles and fields in the quantum domain. Understanding the transient development of particles and their communications is significant for propelling our insight into central material science.

Quantum reproductions offer another apparatus for investigating the worldly complexities of quantum field speculations, possibly prompting forward leaps in how we might interpret the idea of spacetime itself.

The remarkable properties of quantum ensnarement, where particles become interconnected and share data quickly paying little mind to separate, likewise add to the worldly parts of quantum reproductions. The non-neighborhood relationships presented by ensnarement can empower quantum PCs to mimic frameworks with perplexing transient entrapments more productively than old style PCs. This opens up additional opportunities for investigating quantum frameworks with entrapped worldly elements.

While the capability of quantum registering in worldly reproductions is energizing, huge difficulties should be tended to. Quantum frameworks are fragile and powerless to ecological unsettling influences, prompting a peculiarity known as decoherence, where the quantum data is lost. Relieving decoherence is a significant part of creating functional and solid quantum PCs for reproductions with long fleeting elements.

One more test lies in the blunder rates related with quantum calculations. Quantum PCs are innately probabilistic, and blunders can emerge because of variables like blemished doors and associations with the climate. Creating mistake amendment strategies and shortcoming open minded quantum figuring designs is fundamental to guarantee the precision and unwavering quality of quantum reenactments, particularly for applications including many-sided worldly elements.

The acknowledgment of viable quantum registering for fleeting reproductions requires headways in quantum equipment, programming, and blunder remedy techniques. Analysts and specialists are effectively investigating different quantum registering stages, including superconducting qubits, caught particles, and topological qubits, each with its own arrangement of benefits and difficulties. As these advances full grown, the potential for quantum PCs to reform how we might interpret worldly elements turns out to be progressively substantial.

The interdisciplinary idea of quantum processing and its applications in transient recreations additionally features the requirement for cooperation between physicists, scientific experts, PC researchers, and designers. Quantum calculations for worldly reproductions should be intended to successfully saddle the exceptional properties of quantum PCs. This cooperative exertion stretches out to the advancement of quantum programming apparatuses and programming dialects custom fitted for reproducing transient elements on quantum equipment.

Past its applications in logical exploration, quantum processing's effect on transient recreations reaches out to fields, for example, streamlining and AI. Quantum calculations, for example, the quantum strengthening approach, have shown guarantee in taking care of enhancement issues that have fleeting conditions.

Quantum AI calculations, utilizing the standards of quantum parallelism, offer the possibility to deal with worldly information all the more effectively and reveal stowed away examples in enormous datasets.

As quantum PCs progress towards commonsense applications, the investigation of their effect on fleeting reproductions ventures into the domain of fake general insight (AGI). The capacity of quantum PCs to handle data dramatically quicker than old style PCs brings up issues about their job in speeding up the improvement of AGI. The worldly parts of learning and decision-production in AGI frameworks could profit from the remarkable capacities of quantum PCs, introducing another period of clever frameworks with upgraded fleeting handling.

Moral contemplations additionally come to the very front as quantum registering progresses in the domain of transient reenactments. The potential for quantum PCs to tackle complex issues, incorporating those with moral ramifications, brings up issues about mindful and moral use. Guaranteeing that quantum figuring advancements are created and sent in arrangement with moral structures becomes critical as their applications reach out into touchy regions like medication revelation, materials science, and man-made reasoning.

7.3 Speculations on the End of Time

Conjecturing on the finish of time is a scholarly activity that has enraptured the human creative mind for a really long time. It includes examining a definitive destiny of the universe, the idea of time itself, and the existential inquiries that emerge while thinking about the finitude of presence. While such hypotheses frequently span the domains of science, reasoning, and religious philosophy, they stay cryptic, welcoming us to investigate the outskirts of our comprehension.

One road of hypothesis spins around the cosmological idea of the "heat passing" or "entropy demise" of the universe. As indicated by this thought, as the universe grows, matter and energy become progressively scattered, prompting a condition of most extreme entropy where no more energy inclinations exist. In such a situation, all cycles in the universe would stop, and the universe would be in a condition of uniform, perpetual balance.

The intensity passing speculation illustrates a chilly, dull, and dormant universe where time itself appears to lose its significance. It brings up issues about the idea of time past this speculative endpoint - does time endure in a universe without change and elements? The idea challenges our natural comprehension of time as a ceaseless and irreversible stream.

With regards to warm demise, the bolt of time, addressing the imbalance among past and future, takes on an alternate importance. On the off chance that the universe arrives at a condition of most extreme entropy, where confusion is at its pinnacle, the differentiation among past and future could lose its pertinence. Time, as far as we might be concerned, could turn into a component of a universe with dynamic cycles and energy slopes, and industriousness might be dependent upon the circumstances consider such lopsidedness.

Quantum mechanics acquaints one more layer of intricacy with hypotheses on the finish of time. The inborn vulnerability and probabilistic nature of quantum frameworks challenge deterministic perspectives on the universe's destiny. Quantum variances at minuscule scopes could have naturally visible outcomes, prompting the rise of new designs or even the chance of a quantum revival of the universe.

The idea of a cyclic universe, where the finish of one grandiose cycle prompts the introduction of another one, adds a powerful component to these hypotheses. In this situation, the universe goes through an everlasting pattern of development, compression, and resurrection. The cyclic model difficulties the idea of a conclusive finish to time, proposing a recurrent example that rises above the customary direct comprehension of fleeting movement.

String hypothesis, a hypothetical structure in physical science, presents the chance of various aspects past the natural three spatial aspects and once aspect. With regards to string hypothesis, time could be a more complicated and diverse part of the real world. Hypotheses emerge about the potential for universes existing in higher aspects, where time might appear in manners past our ongoing understanding.

Vast peculiarities like dark openings, gravitational singularities, and the idea of spacetime arch contribute extra layers to hypotheses on the finish of time. Dark openings, with their monstrous gravitational force, challenge how we might interpret time by mutilating it in manners that oppose old style material science. The peculiarity inside a dark opening, where gravity turns out to be areas of strength for boundlessly, inquiries regarding the idea of time at such outrageous gravitational circumstances.

The idea of a "major tear" addresses a situation where the development of the universe advances to where worlds, stars, planets, and even iotas are destroyed. In this hypothesis, the finish of time is set apart by the unrestrained extension of the universe, coming full circle in the disintegration, everything being equal. The enormous tear speculation challenges our thoughts of strength and lastingness even with grandiose powers.

Hypotheses on the finish of time likewise cross with philosophical and religious investigations into the idea of presence and reason. In philosophical practices, eschatological convictions frequently consolidate dreams of the finish of time as a zenith of heavenly purposes. The possibility of a last judgment, infinite reestablishment, or the rise of an otherworldly reality adds to the wealth of hypotheses on a definitive destiny of the universe.

Insightfully, contemplations of time's end brief reflections on the importance and meaning of human life inside the transient system. The brevity of life and the possible finitude of the universe bring up significant issues about the motivation behind our undertakings and the quest for information. Guessing on the finish of time turns into a road for examining the human experience inside the immense embroidery of grandiose presence.

The idea of a "reenacted reality" acquaints one more layer with hypotheses on the finish of time. In this situation, the universe and the progression of time could be a recreated develop, similar to a PC program or a refined computer generated simulation. On the off chance that the universe is a reproduction, questions emerge about the idea of the substances or creatures running the reenactment and whether there is a possible end point for the mimicked reality.

The interlacing of logical, philosophical, and religious viewpoints on the finish of time mirrors the intricacy and profundity of these hypotheses. Every road of request adds to a more extensive comprehension of fleeting elements and the idea of the real world. Whether considering the intensity demise, cyclic models, quantum revival, or eschatological convictions, these hypotheses welcome us to defy the secrets that encompass a definitive destiny of the universe and the fleetingness of presence.

Amidst these hypotheses, the job of human organization and the quest for information becomes the dominant focal point. Logical undertakings, from testing the major powers of the universe to creating progressed hypotheses, add to our aggregate comprehension of the universe's fleeting aspects. The steady quest for information permits us to draw in with the significant inquiries that emerge while pondering the finish of time.

The approach of cutting edge cosmic perceptions, space investigation, and hypothetical headways in material science keeps on pushing the limits of our comprehension. The recognition of gravitational waves, the investigation of grandiose microwave foundation radiation, and the investigation of far off cosmic systems give important experiences into the advancement and fleeting elements of the universe.

Mechanical progressions, for example, strong telescopes and molecule gas pedals, empower researchers to peer further into the universe and lead explores that test the essential idea of the real world. The cooperative endeavors of analysts across disciplines add to our capacity to draw in with hypotheses on the finish of time in progressively complex ways.

As we examine these hypotheses, moral contemplations likewise come to the front. The capable utilization of logical information, the moral ramifications of mechanical progressions, and the expected cultural effects of revelations all require cautious thought. The quest for understanding the finish of time ought to be directed by a promise to mindful stewardship of information and a smart assessment of the moral aspects intrinsic in the mission for vast comprehension.

All in all, hypothesizing on the finish of time is a multi-layered try that rises above disciplinary limits. From the domains of physical science and cosmology to reasoning and religious philosophy, these hypotheses welcome us to investigate the secrets of the universe and the idea of transient presence. Whether thinking about the intensity passing, cyclic models, quantum restoration, or eschatological convictions, every road of request adds to a nuanced comprehension of time, presence, and a definitive destiny of the universe. The quest for information and the consideration of these hypotheses address a demonstration of the human limit with respect to interest and a significant commitment with the secrets that encompass us.

7.4 Implications for Philosophy, Religion, and the Human Experience

The investigation of significant logical ideas and mechanical progressions has expansive ramifications for reasoning, religion, and the human experience. As how we might interpret the universe extends and our capacities to control the texture of reality develop, these disciplines are stood up to with new inquiries and difficulties that reshape the manner in which we see the world and our place in it.

Reasoning, as a discipline worried about principal inquiries regarding presence, information, values, reason, psyche, and language, ends up at the junction of logical revelation and mechanical development. The ramifications of advances in fields like quantum mechanics, computerized reasoning, and cosmology force savants to reconsider customary suspicions and dig into the complexities of the real world.

Quantum mechanics, with its inborn vulnerabilities and non-deterministic nature, challenges old style thoughts of causality and determinism. Logicians wrestle with the ramifications of a universe where particles can exist in different states all the while until noticed, bringing up issues about the idea of the real world and the job of perception in forming the world. The translation of quantum mechanics turns into a philosophical riddle, welcoming consideration on the connection among cognizance and the texture of the universe.

The convergence of reasoning and man-made brainpower (computer based intelligence) digs into inquiries of cognizance, mindfulness, and the idea of knowledge. As computer based intelligence frameworks become progressively refined, the well established philosophical investigation into the brain body issue takes on new aspects. The rise of fake general knowledge (AGI), with the potential for mindfulness and independent direction, prompts philosophical reflections on the limits among machine and human cognizance.

Moral contemplations likewise become the dominant focal point in the way of thinking of simulated intelligence. The turn of events and organization of keen machines bring up issues about liability, responsibility, and the moral treatment of computer based intelligence substances. Logicians investigate the ethical elements of making creatures with mental capacities, mulling over the ramifications for human qualities and the moral structures that ought to direct the cooperation among people and savvy machines.

In the domain of reasoning of brain, the union of neuroscience, man-made intelligence, and mental science prompts investigations into the idea of awareness and abstract insight. The investigation of brain organizations, cerebrum PC interfaces, and the potential for mind transferring difficulties conventional dualistic perspectives on the psyche and body. The philosophical investigation of being cognizant in a time of progressing neurotechnology extends how we might interpret the human experience.

Cosmological disclosures and the investigation of the idea of the universe additionally have significant ramifications for philosophical idea. The acknowledgment of the immensity and intricacy of the universe prompts reflections on our spot in the universe and the existential inquiries that emerge from examining the hugeness of reality. The infinite viewpoint welcomes savants to reevaluate human-centric perspectives and think about the ramifications of our enormous irrelevance.

Religion, with its rich embroidery of convictions, customs, and moral lessons, experiences the two difficulties and open doors despite logical and mechanical headways. The revelations in cosmology, like the comprehension of the age and beginning of the universe, brief religious reflections on the idea of creation and the connection between the heavenly and the universe.

The ramifications of quantum mechanics acquaint a layer of intricacy with philosophical conversations about divine all-knowingness and transcendence. The inborn vulnerabilities and indeterminacies at the quantum level bring up issues about whether a divinity could have total information on the universe's future states. Philosophical philosophy draws in with these inquiries, investigating the similarity between quantum indeterminacy and religious ideas of heavenly premonition.

The approach of artificial intelligence likewise brings up philosophical issues about the idea of cognizance, choice, and the spirit. The making of savvy machines prompts strict reflections on the uniqueness of individuals and the moral contemplations encompassing the possible improvement of cognizant elements. Philosophical conversations wrestle with the ramifications of instilling machines with mental capacities and the religious status of computer based intelligence substances.

Cosmological ideas, like the intensity demise of the universe or the possible repeating nature of infinite development, challenge strict stories about a definitive destiny of the universe. Religious reflections on eschatology, the investigation of end times, should draw in with logical experiences into the likely finitude or everlasting repeat of the universe. The powerful transaction between logical revelations and philosophical tenets shapes the manner in which strict customs decipher and answer the secrets of the universe.

In the domain of reasoning and religion, the human experience is unpredictably woven into the texture of request. Inquiries regarding the significance of life, the idea of profound quality, and the motivation behind presence penetrate the two disciplines. The effect of logical and innovative progressions on the human experience is significant, reshaping how we might interpret character, organization, and the moral elements of our decisions.

Progresses in biotechnology and the potential for human improvement brief philosophical and moral reflections on the idea of human character and the limits of personhood. Inquiries concerning the moral utilization of hereditary designing, mental improvement, and the combination of innovation into the human body challenge customary perspectives on being human. Philosophical investigations into the moral elements of human upgrade dig into issues of independence, equity, and the protection of human poise.

The crossing point of neuroscience and reasoning of brain investigates the complexities of cognizance and mindfulness. As how we might interpret the brain corresponds of cognizance propels, rationalists wrestle with inquiries regarding the idea of abstract insight and the connection between the cerebrum and the psyche. The investigation of adjusted conditions of awareness, for example, those instigated by hallucinogenics or neurotechnological mediations, further extends how we might interpret the human experience.

Moral contemplations in the domain of human improvement stretch out to inquiries of civil rights and the potential for fueling existing disparities. Philosophical reflections on the circulation of mechanical advantages, admittance to improvement advances, and the moral obligations of those forming innovative progressions become pivotal in guaranteeing that the human experience is enhanced as opposed to divided by these developments.

The human experience is additionally significantly formed by the interconnectedness worked with by data advances. The approach of the web, online entertainment, and computer generated reality acquaints new aspects with human correspondence, social communication, and the arrangement of networks. Philosophical investigations into the morals of online way of behaving, advanced character, and the effect of virtual spaces on human connections investigate the developing scene of the human involvement with the computerized age.

Existential inquiries regarding the importance of life and the quest for satisfaction resound in both way of thinking and the human experience. The mission for a significant life, informed by private qualities, social impacts, and philosophical reflections, rises above disciplinary limits. As people explore the intricacies of presence, the exchange between philosophical structures and lived encounters shapes the accounts of direction and satisfaction.

The ramifications of logical and innovative progressions on the human experience reach out to inquiries of ecological morals and our obligation as stewards of the planet. The biological difficulties presented by environmental change, asset exhaustion, and ecological corruption brief

philosophical reflections on humankind's relationship with the normal world. Moral contemplations about manageable practices, preservation endeavors, and the effect of human exercises on biological systems become essential to forming an agreeable and mindful human experience.

All in all, the ramifications for reasoning, religion, and the human involvement with the essence of logical and mechanical progressions are immense and unpredictable. As we explore the boondocks of information and wrestle with the intricacies of presence, these disciplines become fundamental focal points through which we decipher, question, and figure out the advancing idea of the real world and our place inside it. The powerful exchange between logical request, philosophical reflection, strict consideration, and the lived encounters of people makes a rich embroidery that keeps on unfurling as we on the whole investigate the secrets of the universe and the profundities of the human spirit.